PENGUIN BOOKS

FREEDOM EVOLVES

'Outstandingly good . . . should make most readers think afresh about their assumptions' Michael Rutter, *The Times Higher Education Supplement*

'Provocative and stimulating . . . written with admirable clarity . . . this important book, remarkably, throws new light on a well-worn topic' Patrick Masterson, *Irish Times*

'Daniel Dennett's new book combines, once again, original philosophical thinking, marvellously vivid prose and extraordinarily lucid argumentation. *Freedom Evolves* does what I would have thought impossible: it says something new about free will and determinism' Richard Rorty

'In a series of sparkling books, Dennett has taken on really big issues, made them clear, dealt with them seriously and given us much on which to reflect and (most important) with which to disagree . . . Dennett is crisp and critically insightful on all sorts of flabby presuppositions . . . This is a man who truly loves science' Michael Ruse, *Washington Post*

'Engaging and very smart . . . lively and highly original . . . As in his previous books, he has rendered the difficult comprehensible' Melvin Konner, *Nature*

'Dennett is not just a philosophy professor, but a genuine philosopher, much to our benefit. In *Freedom Evolves*, he takes on the question of free will and determinism, one of the oldest and most intransigent of conundrums, transporting the discussion where it belongs, into the realm of Darwinian thought' David Barash, *Human Nature Review*

ABOUT THE AUTHOR

Daniel C. Dennett is the author of *Brainstorms, Brainchildren, Elbow Room, Consciousness Explained* and *Darwin's Dangerous Idea*. He is currently the Distinguished Arts and Sciences Professor and Director of the Center for Cognitive Studies at Tuft's University. He lives in North Andover, Massachusetts.

FREEDOM EVOLVES

Daniel C. Dennett

PENGUIN BOOKS

PENGUIN BOOKS

Published by the Penguin Group
Penguin Books Ltd, 80 Strand, London WC2R 0RL, England
Penguin Group (USA), Inc., 375 Hudson Street, New York, New York 10014, USA
Penguin Books Australia Ltd, 250 Camberwell Road,
Camberwell, Victoria 3124, Australia
Penguin Books Canada Ltd, 10 Alcorn Avenue, Toronto, Ontario, Canada M4V 3B2
Penguin Books India (P) Ltd, 11 Community Centre,
Panchsheel Park, New Delhi – 110 017, India
Penguin Books (NZ) Ltd, Cnr Rosedale and Airborne Roads,
Albany, Auckland, New Zealand
Penguin Books (South Africa) (Pty) Ltd, 24 Sturdee Avenue,
Rosebank 2196, South Africa

Penguin Books Ltd, Registered Offices: 80 Strand, London WC2R 0RL, England

www.penguin.com

First published in the United States of America by Viking Penguin, a division of Penguin
Putnam Inc. 2003
First published in Great Britain by Allen Lane 2003
Published in Penguin Books 2004

6

Figures 7.1, 7.2 and 7.3 reproduced from *Breakdown of Will* by George Ainslie (Cambridge
University Press, 2001). Used by permission of the author. Figures 8.1 and 8.2 reproduced
from 'Do We Have Free Will?' by Benjamin Libet from *The Volitional Brain: Towards a
Neuroscience of Free Will* edited by Benjamin Libet, Keith Sutherland, and Anthony Freeman
(Imprint Academic, 1999). Used by permission of the author

Set by Rowland Phototypesetting Ltd, Bury St Edmunds, Suffolk
Printed in England by Clays Ltd, St Ives plc

ISBN-13: 978–0–140–28389–1
ISBN-10: 0–140–28389–7

For my family:
Susan, Peter, Andrea, Nathan, and Brandon

CONTENTS

Chapter 8

ARE YOU OUT OF THE LOOP?

Chapter 9

BOOTSTRAPPING OURSELVES FREE

Chapter 10

THE FUTURE OF HUMAN FREEDOM

PREFACE

How long have I been working on this book? As I was doing the final editing, several people asked me and I didn't know how to answer: five years or thirty years? Thirty years is closer to the truth, I think, since it was roughly that long ago that I began thinking in earnest about the topics, reading the relevant literature, drafting arguments, making lists of further books and articles to read, plotting strategy and structure, and engaging in debate and discussion. From the thirty-year bird's-eye view, my 1984 book, *Elbow Room: The Varieties of Free Will Worth Wanting,* counts as a pilot project. It relied heavily on a simple ten-page sketch of the evolution of consciousness (pp. 34–43) accompanied by two promissory notes: Owed to the skeptical reader were properly detailed accounts of both consciousness and evolution. It took me a dozen years to keep those promises, in *Consciousness Explained* (Dennett 1991A) and *Darwin's Dangerous Idea* (Dennett 1995). During that time I continued to notice instances of the pattern that had inspired and shaped *Elbow Room:* the hidden agenda that tends to distort theorizing in all the social sciences and life sciences. People working in quite different fields with different methodologies and research agendas nevertheless often shared a veiled antipathy, trying to keep their distance from the implications of two ideas: Our minds are just what our brains non-miraculously do, and the talents of our brains had to evolve like every other marvel of nature. Their effort to keep this vision at bay was bogging down their thinking, lending spurious allure to dubious brands of absolutism and encouraging them to see small, bridgeable gaps as yawning chasms. The aim of this book is to expose the misbegotten defensive edifices people have constructed in response to this fear, dismantle them, and replace them with better foundations for the things we hold dear.

In 2001, the home stretch, I had superb help, both institutional and personal. My academic home all these years, Tufts University, gave me a sabbatical semester. Once again the Rockefeller Foundation's Villa Serbelloni in Bellagio provided the perfect setting for writing, and first drafts of half the chapters emerged from an intense month of work, illuminated by discussions and suggestions from the other residents, especially Sheldon Siegel, Bernard Gross, Rita Charon, Frank Levy, Evelyn Fox Keller, Julie Barmazel, Mary Childers, and Gerald Postema. Then Sandro Nannini and his students and colleagues at the University of Siena provided a vigorous and knowledgeable audience for the debut of some of the book's central arguments.

    In April I took up residence as Leverhulme Visiting Professor at the London School of Economics, where I presented the first seven chapters as weekly public lectures followed by seminars the next day, supplemented by many informal discussions both at LSE and on several visits to Oxford. John Worrall, Nick Humphrey, Richard Dawkins, John Maynard Smith, Matteo Mameli, Nicholas Maxwell, Oliver Curry, Helena Cronin, K. M. Dowding, Susan Blackmore, Antti Saaristo, Janne Mantykoski, Valerie Porter, Isabel Gois, and Katrina Sifferd all provided valuable reactions, rebuttals, refinements, and suggestions.

    To Christopher Taylor I owe much of the perspective-shifting thinking that is incorporated in our jointly authored paper and featured in Chapter 3, as well as many penetrating suggestions on the drafts of other chapters. To David Benedictus, an extraordinary writer and friend for even more than thirty years, I owe thanks for a different kind of perspective-shifting that eventually led to the book's title. Robert Kane and Daniel Wegner, whose books get criticized here (constructively, I hope!), were very generous with their comments on my treatment of their brainchildren. Other friends and colleagues who have read large portions of various drafts and provided advice both editorial and substantive are, in alphabetical order, Andrew Brook, Michael

Cappucci, Tom Clark, Mary Coleman, Bo Dahlbom, Gary Drescher, Paulina Essunger, Marc Hauser, Erin Kelly, Kathrin Koslicki, Paul Oppenheim, Will Provine, Peter Reid, Don Ross, Scott Sehon, Mitch Silver, Elliott Sober, Matthew Stuart, Peter Suber, Jackie Taylor, and Steve White.

I was able to continue my tradition of playing Tom Sawyer and the whitewashed fence with the penultimate draft of this book, which was intelligently swarmed over and taken to pieces by a large and opinionated horde of students and auditors, undergraduate and graduate, in my fall seminar. James Arinello, David Baptista, Matt Bedoukian, Lindsay Beyerstein, Cinnamon Bidwell, Robert Briscoe, Hector Canseco, Russell Capone, Regina Chouza, Ashley de Marchena, Janelle DeWitt, Jason Disterhoft, Jennifer Durette, Gabrielle Jackson, Ann J. Johnson, Sarah Jurgensen, Tomasz Kozyra, Marcy Latta, Ryan Long, Gabriel Love, Carey Morewedge, Brett Mulder, Cathy Muller, Sebastian S. Reeve, Daniel Rosenberg, Amber Ross, George A. Samuel, Derek Sanger, Shorena Shaverdashvili, Mark Shwayder, Andrew Silver, Naomi Sleeper, Sara Smollett, Rodrigo Vanegas, Nick Wakeman, Jason Walker, and Robert Woo all provided comments, leading to dozens of improvements. The errors and shortcomings that remain are not their fault; they did their best to set me straight.

I am grateful to Craig Garcia and Durwood Marshall for the original figures; to Teresa Salvato and Gabriel Love in the Center for Cognitive Studies for uncounted library runs and clerical help with the preparation of the many drafts of the manuscript; and to the Collegium Budapest, which provided an intellectually stimulating and gracious home away from home during the final copyediting and revisions.

Last, and most important, once again, thanks and love to my wife, Susan, for more than forty years of advice, love, and support.

DANIEL DENNETT
JUNE 20, 2002

FREEDOM
EVOLVES

Chapter 1

NATURAL FREEDOM

One widespread tradition has it that we human beings are responsible agents, captains of our fate, *because* what we really are are *souls*, immaterial and immortal clumps of Godstuff that inhabit and control our material bodies rather like spectral puppeteers. It is our souls that are the source of all meaning, and the locus of all our suffering, our joy, our glory and shame. But this idea of immaterial souls, capable of defying the laws of physics, has outlived its credibility thanks to the advance of the natural sciences. Many people think the implications of this are dreadful: We don't really have "free will" and nothing really matters. The aim of this book is to show why they are wrong.

Learning What We Are

Sì, abbiamo un anima. Ma è fatta di tanti piccoli robot.
Yes, we have a soul. But it's made of lots of tiny robots.

—Giulio Giorelli

We don't have to have immaterial souls of the old-fashioned sort in order to live up to our hopes; our aspirations as moral beings whose acts and lives matter do not depend *at all* on our having minds that obey a different physics from the rest of nature. The self-understanding we can gain from science can help us put our moral lives on a new and better foundation, and once we understand what our freedom consists in, we will be much better prepared to protect it against the genuine threats that are so regularly misidentified.

A student of mine who went into the Peace Corps to avoid serving in the Vietnam War later told me about his efforts on behalf of a tribe living deep in the Brazilian forest. I asked him if he had been required to tell them about the conflict between the USA and the USSR. Not at all, he replied. There would have been no point in it. They had never heard of either America or the Soviet Union. In fact, they had never even heard of Brazil! It was still possible in the 1960s for a human being to live in a nation, and be subject to its laws, without the slightest knowledge of that fact. If we find this astonishing, it is because we human beings, unlike all other species on the planet, are knowers. We are the only ones who have figured out what we are, and where we are, in this great universe. And we're even beginning to figure out how we got here.

These quite recent discoveries about who we are and how we got here are unnerving, to say the least. What you are is an assemblage of roughly a hundred trillion cells, of thousands of different sorts. The bulk of these cells are "daughters" of the egg cell and sperm cell whose union started you, but they are actually outnumbered by the trillions of bacterial hitchhikers from thousands of different lineages stowed away in your body (Hooper et al. 1998). Each of your host cells is a mindless mechanism, a largely autonomous micro-robot. It is no more conscious than your bacterial guests are. Not a single one of the cells that compose you knows who you are, or cares.

Each trillion-robot team is gathered together in a breathtakingly efficient regime that has no dictator but manages to keep itself organized to repel outsiders, banish the weak, enforce iron rules of discipline—and serve as the headquarters of one conscious self, one mind. These communities of cells are fascistic in the extreme, but *your* interests and values have little or nothing to do with the limited goals of the cells that compose you—fortunately. Some people are gentle and generous, others are ruthless; some are pornographers and others devote their lives to the service of God. It has been tempting over the ages to imagine that these striking differences must be due to the special features of some *extra* thing (a soul) installed somehow in the bodily headquarters. We now know that tempting as this idea still is, it is not supported in the slightest by anything we have learned about our biology in general and our brains in particular. The more we learn about how we have evolved, and how our brains work, the more certain we are becoming that there is no such extra ingredient. We are each *made of* mindless robots and nothing else, no non-physical, non-robotic ingredients at all. The differences among people are all due to the way their particular robotic teams are put together, over a lifetime of growth and experience. The difference between speaking French and speaking Chinese is a difference in the organization of the working parts, and so are all the other differences of knowledge and personality.

Since I am conscious and you are conscious, we must have conscious selves that are *somehow* composed of these strange little parts. How can this be? To see how such an extraordinary composition job could be accomplished, we need to look at the history of the design processes that did all the work, the evolution of human consciousness. We also need to see how these souls made of cellular robots actually do endow us with the important powers and resultant obligations that traditional immaterial souls were supposed to endow us with (by unspecified magic). Trading in a supernatural soul for a natural soul— is this a good bargain? What do we give up and what do we gain? People jump to fearful conclusions about this that are hugely mistaken. I propose to prove this by tracing the growth of *freedom* on our planet from its earliest beginnings at the dawn of life. What kinds of freedom? Different kinds will emerge as the story unfolds.

Four and a half billion years ago, the planet Earth was formed, and it was utterly without life. And so it stayed for perhaps half a billion years, until the first simple life-forms emerged, and then for the next three billion years or so, the planet's oceans teemed with life, but it was all blind and deaf. Simple cells multiplied, engulfing each other, exploiting each other in a thousand ways, but oblivious to the world beyond their membranes. Then finally much larger, more complex cells evolved—eukaryotes—still clueless and robotic, but with enough internal machinery to begin to specialize. So it continued for a few hundred million more years, the time it took for the algorithms of evolution to stumble upon good ways for these cells and their daughters and granddaughters to band together into multicellular organisms composed of millions, billions, and (eventually) trillions of cells, each doing its particular mechanical routine, but now yoked into specialized service, as part of an eye or an ear or a lung or a kidney. These organisms (not the individual team members composing them) had become *long-distance* knowers, able to spy supper trying to appear inconspicuous in the middle distance, able to hear danger threatening from afar. But still, even these whole organisms knew not what they were. Their instincts guaranteed that they tried to mate with the right sorts, and flock with the right sorts, but just as those Brazilians didn't know they were Brazilians, no bison has ever known it's a bison.[1]

In just one species, our species, a new trick evolved: language. It has provided us a broad highway of knowledge-sharing, on every topic. Conversation unites us, in spite of our different languages. We can all know quite a lot about what it is like to be a Vietnamese fisherman or a Bulgarian taxi driver, an eighty-year-old nun or a five-year-old boy blind from birth, a chess master or a prostitute. No matter how different from one another we people are, scattered around the globe, we can explore our differences and communicate about them. No matter how similar to one another bison are, standing shoulder to shoulder in a herd, they cannot know much of anything about their similarities, let alone their differences, because they can't compare notes. They can have similar experiences, side by side, but they really can't share experiences the way we do.

Even in our species, it has taken thousands of years of communication for us to begin to find the keys to our own identities. It has been only a few hundred years that we've known that we are mammals, and only a few decades that we've understood in considerable detail how we have evolved, along with all other living things, from those simple beginnings. We are outnumbered on this planet by our distant cousins, the ants, and outweighed by yet more distant relatives, the bacteria. Though we are in the minority, our capacity for long-distance knowledge gives us powers that dwarf the powers of all the rest of the life on the planet. Now, for the first time in its billions of years of history, our planet is protected by far-seeing sentinels, able to anticipate danger from the distant future—a comet on a collision course, or global warming—and devise schemes for doing something about it. The planet has finally grown its own nervous system: us.

---

1. In general, nature operates on a version of the Need to Know Principle made famous in the world of espionage: Bison don't need to know that they are ungulates within the class Mammalia—there is nothing they could do with that information, being bison; the Brazilians didn't (yet) need to know much about the larger environment of which their intimately known jungle environment formed a part, but the Brazilians, being human beings, could almost effortlessly extend their epistemic horizons as soon as they needed to know. I am sure they know it now.

We may not be up to the job. We may destroy the planet instead of saving it, largely because we are such free-thinking, creative, unruly explorers and adventurers, so unlike the trillions of slavish workers that compose us. Brains are for anticipating the future, so that timely steps can be taken in better directions, but even the smartest of beasts have very limited time horizons, and little if any ability to imagine alternative worlds. We human beings, in contrast, have discovered the mixed blessing of being able to think even about our own deaths and beyond. A huge portion of our energy expenditure over the last ten thousand years has been devoted to assuaging the concerns provoked by this unsettling new vista that we alone have.

If you burn more calories than you take in, you soon die. If you find some tricks that provide you a surplus of calories, what might you spend them on? You *might* devote person-centuries of labor to building temples and tombs and sacrificial pyres on which you destroy some of your most precious possessions—and even some of your very own children. Why on earth would you want to do that? These strange and awful expenditures give us clues about some of the hidden costs of our heightened powers of imagination. We did not come by our knowledge painlessly.

Now what will we do with our knowledge? The birth pangs of our discoveries have not subsided. Many are afraid that learning too much about what we are—trading in mystery for mechanisms—will impoverish our vision of human possibility. This fear is understandable, but if we really were in danger of learning too much, wouldn't those on the cutting edge be showing signs of discomfort? Look around at those who are participating in this quest for further scientific knowledge and eagerly digesting the new discoveries; they are manifestly not short on optimism, moral conviction, engagement in life, commitment to society. In fact, if you want to find anxiety, despair, and anomie among intellectuals today, look to the recently fashionable tribe of postmodernists, who like to claim that modern science is just another in a long line of myths, its institutions and expensive apparatus just the rituals and accoutrements of yet another religion. That intelligent people can take this seriously is a testimony to the power that fearful thinking still has, in spite of our advances in self-knowledge. The postmodernists are right that science is just one of the things we might want to spend our extra calories on. The fact that science has been a major source of the efficiencies that created those extra calories does not entitle it to any particular share of the wealth it has created. But it should still be obvious that the innovations of science—not just its microscopes and telescopes and computers, but its commitment to reason and evidence—are the new sense organs of our species, enabling us to answer questions, solve mysteries, and anticipate the future in ways no earlier human institutions can approach.

The more we learn about what we are, the more options we will discern about what to try to become. Americans have long honored the "self-made man," but now that we are actually learning enough to be able to remake ourselves into something new, many flinch. Many would apparently rather bumble around with their eyes closed, trusting in tradition, than look around to see what's about to happen. Yes, it is unnerving; yes, it can be scary. After all, there are entirely new mistakes we are now empowered to make for the first time. But it's the beginning of a great new adventure for our knowing species. And it's much more exciting, as well as safer, if we open our eyes.

I Am Who I Am

I read in the newspaper recently about a young father who forgot to drop off his infant daughter at the day-care center on his way to work. She spent the day locked in his car in a hot parking lot, and in the evening on his way home when he stopped at the day-care center to pick her up, he was told, "You didn't drop her off today." He rushed out to his car to find her still strapped into her little car seat in the back, dead. If you can bear it, put yourself in this man's shoes. When I do, I shudder; my heart aches at the thought of the unspeakable shame, the self-loathing, the regret beyond regret that this man must now be suffering. And as one who is notoriously absentminded, who readily gets lost in his own thoughts, I find it even more unsettling to ask myself: Could I ever do anything like that? Could I be that negligent with the life of a child in my care? I replay the scene with many variations, imagining distractions—a fire engine racing by just as I am about to turn off to the day-care center, something on the radio reminding me of a problem I have to solve that day, and later, in the parking lot, a friend asking me for help as I get out of my car, or perhaps I drop some papers on the ground and have to pick them up. Could a series of such distractors pile up and bury my overriding project of getting my daughter safely to day care? Could I be so unlucky as to blunder into a situation where events conspired to bring out the very worst in me, exposing my weakness, and leading me down this despicable path? I am so thankful that nothing like this has yet confronted me, because I do not know that there are no circumstances in which I could do what this man did. Such things happen all the time. I know nothing more about this young father. It is conceivable that he is a callous and irresponsible human being, a villain who deserves to be despised by us all. But it is also conceivable that he's basically a good person, a victim of cosmic bad luck. And, of course, the better person he is, the greater his remorse must now be. He must wonder if there is any honorable way to go on living. "I'm the guy who forgot his baby daughter and let her bake to death in his locked car. That's who I am."

Each of us is who he is, warts and all. I can't be a champion golfer or a concert pianist or a quantum physicist. I can live with that. That's part of who I am. Can I break 90 on the golf course, or ever play that Bach fugue through from beginning to end without any mistakes? I can try, it seems, but if I never succeed, will it have been the case that I never could have succeeded, not really? "Be all that you can be!"—a thrilling recruiting slogan for the U.S. Army, but does it conceal a mocking tautology? Aren't we all, automatically, all that we can be? "Hey, I'm an undisciplined, ill-educated, overweight couch potato who apparently doesn't have the gumption to join the army. I already *am* all that *I* can be! I am who I am." Is this fellow deluding himself out of a better life, or has he seen to the heart of the matter? Is there a legitimate sense in which although I really and truly *can't* be a champion golfer, I really and truly *can* break 90? Can any of us ever do anything other than what we end up doing? If not, what's the point of trying? Indeed, what's the point of anything?

What we want to be true, one way or another, is that there is a point. And for several millennia we've struggled with a family of arguments that imply that there may not be any point, because if the world is the way science tells us the world is, there is no room for our strivings and yearnings. The ancient Greek atomists had no sooner dreamed up the brilliant idea that the world was composed of myriad tiny par-

ticles bouncing off each other than they hit upon the corollary that in
that case, every event, including our every heartbeat, fib, and private
self-admonition, unfolds according to laws of nature that *determine* what
happens next down to the finest details and thus provide no options, no
real choice points, no opportunities for things to be one way rather than
another. If *determinism* is true, although there may well *seem* to be a
point, this is an illusion. Indeed, we may well be determined to go on
thinking that there is a point, but if so, we will be wrong. So it has often
seemed. Naturally this has fueled the hope that the laws of nature are
not deterministic after all. The first attempt to soften the blow of atom-
ism was by Epicurus and his followers, who proposed that a *random
swerve* in the trajectories of some of those atoms might provide the
elbow room for free choice, but since wishful thinking was their only
grounds for postulating this random swerve, it was met from the outset
with deserved skepticism. But don't give up hope. Quantum physics to
the rescue! When we learn that down in the strange world of subatomic
physics, different rules apply, indeterministic rules, this quite appropriately gives rise to a new quest: showing how we can harness this quantum indeterminism to open up a model of a human being as a striver
with genuine opportunities, capable of making truly free decisions.

       This is such a perennially attractive option that it needs to be
given careful, sympathetic review, and in Chapter 4 it will get one, but
I will argue, as many before me have argued, that it just won't work.
As William James put it almost a century ago,

> If a "free" act be a sheer novelty, that comes not *from* me, the pre-
> vious me, but *ex nihilo,* and simply tacks itself on to me, how can
> *I,* the previous I, be responsible? How can I have any permanent
> *character* that will stand still long enough for praise or blame to be
> awarded? (James 1907, p. 53)

How indeed? I advise my students to be on the lookout for rhetorical
questions, which typically mark the weakest link in any defense. A
rhetorical question implies a reductio ad absurdum argument too obvious to need spelling out, the perfect hiding place for an unexamined
assumption that might better be explicitly denied. One can often
embarrass the asker of a rhetorical question by simply trying to answer
it: "*I'll* show you how!" We will consider just such an attempt in Chapter 4, and we will see that James's challenge can in fact be met in most

regards. He overstates the case in several ways when he concludes: "The chaplet of my days tumbles into a case of disconnected beads as soon as the thread of inner necessity is drawn out by the preposterous indeterminist doctrine." Indeterminism is not preposterous, but it is also no help to those who crave free will, and our examination will reveal some surprises about how our imaginations have been deflected in the search for a solution to the problem of free will.

The Air We Breathe

People are surprisingly good at distracting themselves from ominous prospects, and nowhere have they done a better job of diverting their attention from the real problem than on the issue of free will. The classical problem of free will, defined and endorsed by centuries of work by philosophers, theologians, and scientists, asks whether the world is so constituted as to permit us to make genuinely free, responsible decisions. The answer depends, it has always seemed, on basic, eternal facts—the fundamental laws of physics (whatever they turn out to be) and definitional truths about the nature of matter, time, and causation, and equally fundamental definitional truths about the nature of our minds, such as the fact that a stone or a sunflower couldn't possibly have free will—only something with a mind is even a candidate for this blessing, whatever it is. I will try to show that this traditional problem of free will is, in spite of its pedigree, a distractor, a puzzle of no real importance that draws our attention away from some neighboring concerns that truly matter, that *ought* to keep us awake nights worrying. These concerns typically get set aside as empirical complications that muddy the metaphysical water, but I want to resist that deflection and promote these tangential issues into the main topic. The genuine threat, the submerged source of the anxiety that makes the free will topic such a perennial riveter of attention in philosophy courses, arises from a set of facts about the human situation that are empirical, and even, in one sense, political: They are sensitive to human attitude. It really makes a difference what we think about them.

We live our lives against a background of facts, some of them variable and some of them rock solid. Some of the stability comes from fundamental physical facts: The law of gravity will never let us down (it will always pull us down, so long as we stay on Earth), and we can rely on the speed of light staying constant in all our endeavors.[2] Some of the stability comes from even more fundamental, *meta*physical facts: 2 + 2 will always add up to 4, the Pythagorean theorem will hold, and if A = B, whatever is true of A is true of B and vice versa. The idea that we have free will is another background condition for our whole way of thinking about our lives. We count on it; we count on people "having free will" the same way we count on them falling when pushed off cliffs and needing food and water to live, but it is neither a metaphysical background condition nor a fundamental physical condition. Free will is like the air we breathe, and it is present almost everywhere we want to go, but it is not only not eternal, it evolved, and is still evolving. The atmosphere of our planet evolved over hundreds of millions of years as a product of the activities of simple early life-forms, and it continues to evolve today in response to the activities of the billions of more complex life-forms it made possible. The atmosphere of free will is another sort of environment. It is the enveloping, enabling, life-shaping, *conceptual* atmosphere of intentional action, planning and hoping and promising— and blaming, resenting, punishing, and honoring. We all grow up in this conceptual atmosphere, and we learn to conduct our lives in the terms it provides. It *appears to be* a stable and ahistorical construct, as eternal and unchanging as arithmetic, but it is not. It evolved as a recent product of human interactions, and some of the sorts of human activity it first made possible on this planet may also threaten to disrupt its future stability, or even hasten its demise. Our planet's atmosphere is not guaranteed to last forever, and neither is our free will.

We are already taking steps to prevent the deterioration of the air we breathe. They may be too little too late. We can imagine devising technological innovations (giant air-conditioning domes, *terra*-lungs?) that would permit us to live on without the natural atmosphere. Life would be very different, and very difficult, but it might still be life worth living. What happens, though, when we try to imagine living in a world without the atmosphere of free will? It might be life, but would it be *us*? Would life be worth living if we lost our belief in our

2. Or nearly constant. Some recent, and controversial, evidence from the far reaches of space suggests to some scientists that there *might* be some change in the speed of light over cosmological time periods.

own capacity to make free, responsible decisions? And is the ubiqui-tous atmosphere of free will in which we live and act not a *fact* at all, but just a facade of some sort, a mass hallucination?

There are those who say that free will has always been an illu-sion, a pre-scientific dream from which we are now awakening. We've never *really* had free will, and never could have had it. Thinking we've had free will has been, at best, a life-shaping and even life-enhancing ideology, but we can learn to live without it. Some people claim already to have done so, but what they mean by this is not clear. Some of them insist that although free will is an illusion, this discovery has no signifi-cant bearing on how they think about their lives, their hopes and plans and fears, but they do not bother elaborating on this curious separation of issues. Others excuse the vestiges of the creed that persist in their ways of speaking and thinking as largely harmless habits they haven't bothered to outgrow, or as diplomatic concessions to the traditional manners of the less advanced thinkers around them. They go along with the crowd, accepting "responsibility" for "decisions" that were not really free, blam-ing and praising others while keeping their fingers crossed, knowing that deep down, nobody ever deserves anything because everything that hap-pens just spins out of the vast network of mindless causes that prevents anything from meaning anything, in the final analysis.

Are these self-styled dis-illusioned ones making a big mistake? Are they discarding a precious perspective for no good reason, dazzled by a misreading of science into accepting a diminished self-image? And does it matter one way or the other? It is tempting to dismiss the ques-tion of free will as just another philosophers' puzzle, an artificial stumper created by a conspiracy of ingenious definitions. Do you have free will? "Well," says the philosopher, lighting his pipe, "it all depends on what you mean by free will; now, on the one hand, if you adopt a *compatibilist* definition of free will, then . . ." (and we're off to the races). To see that the stakes are higher, that the issues really do matter, it helps to make them personal. Reflect, then, on your adult life and pick a truly bad moment, as bad a moment as you can bear to contemplate in suffocating detail. (Or, if that is too painful, just try putting yourself for a moment in the young father's shoes.) So fix the terrible act in your mind; you did it. If only you hadn't done it!

Now, so what? In the larger scheme of things, what is the meaning of your regret? Does it count for anything, or is it just a sort of involuntary hiccup, a meaningless spasm provoked by a meaningless world? Do we live in a universe in which striving and hoping, regretting, blaming, promising, trying to do better, condemning and praising make sense? Or are they all part of a vast illusion, honored by tradition but overdue for exposure?

Some people—you may be one—may be momentarily comforted to conclude that they don't have free will, and that none of it matters, neither the shameful violations nor the glorious triumphs; it's all just the unwinding of pointless clockwork. This may seem to them like a great relief at first, but then they may reflect, with irritation, that they nevertheless cannot help caring, cannot keep themselves from worrying, striving, hoping—and then go on to reflect that moreover they can't help being irritated by their incessant desire to care, and so forth, a downward spiral into the motivational equivalent of the Heat Death of the universe: Nothing moves, nothing matters, nothing.

Other people—you may be one—are sure they have free will. They don't just strive; they embrace their own strivings, defying their so-called fate. They envision possibilities, trying to make the most of golden opportunities and thrilling in narrow escapes from disaster. They take themselves to be in charge of their own lives and responsible for their own deeds.

There might, it seems, be two kinds of people: those who believe that they don't have free will (even if they can't help acting most of the time as if they believed they did), and those who believe they do have free will (even if this is an illusion). Which group are you in? Which group is better off, happier? But, finally, which group is *right?* Are those in the first group the undeluded ones, seeing through the grand illusion at least in their reflective moments? Or are they the ones who are missing the point, victimized by some cognitive illusions that tempt them to turn their backs on the truth, disabling themselves by discarding the very idea that gives life its meaning? (Too bad, but maybe they can't help it. Maybe they are *determined* by their past, their genes, their upbringing, their education, to reject the idea of free will! As the comedian Emo Phillips has quipped: "I'm not a fatalist, but even if I were, what could I do about it?")

This raises what may be yet another possibility. Perhaps there are two kinds of normal people (setting aside those who are truly disabled and could not possibly have free will because they are comatose *what you think matters*

or demented): There are those who don't believe in free will and *thereby* don't have free will, and there are those who do believe in free will and *thereby* actually have free will. Might something like "the power of positive thinking" actually be great enough to make the crucial difference? This might not give much solace, since it could still be, it seems, that it's just the luck of the draw which group you're in, for better or for worse. Might you switch groups? Might you want to? It is fiendishly hard to keep this curious aspect of free will in focus. If it is a brute metaphysical fact that people do (or don't) have free will, then this cannot be influenced by "majority rule" or anything of that kind, and your only option (option?—do we really have *options?*) is whether or not you want to know whatever the metaphysical truth is. But people often talk and write as if they were, in effect, *campaigning for* the belief in free will, as if free will (not just the belief in free will) were a political condition that might be under threat, might spread or go extinct as a result of what people came to believe. Is free will like democracy, perhaps? What is the relation between political freedom and (*metaphysical,* for want of a better word) free will?

In the rest of the book, my task will be to bring this churning of perspectives to a halt and provide a unified, stable, empirically well-grounded, coherent view of human free will, and you already know the conclusion I will reach: Free will is real, but it is not a preexisting feature of our existence, like the law of gravity. It is also not what tradition declares it to be: a God-like power to exempt oneself from the causal fabric of the physical world. It is an evolved creation of human activity and beliefs, and it is just as real as such other human creations as music and money. And even more valuable. From this evolutionary perspective, the traditional problem of free will can be broken into some rather unusual fragments, each of some value in illuminating the *serious* problems of free will, but we can undertake this reexamination only after we have corrected the misdirection implicit in their traditional settings.

Dumbo's Magic Feather and the Peril of Paulina

In Walt Disney's classic animated film *Dumbo,* about the little elephant who learns to spread his giant ears and fly, there is a pivotal scene in which a dubious—indeed terrified—Dumbo is being cajoled by his friends, the crows, to leap off a cliff into the air, proving to himself that he can fly. One of the crows has a bright idea. When Dumbo isn't looking he plucks a tail feather from one of his kind and then ceremonially hands it to Dumbo, announcing that it is a magic feather: So long as Dumbo clutches it in his trunk, he can fly! The scene is presented with masterful economy. No explanation is provided, since even small children get the point without being told: The feather isn't really magic; it's a prosthetic device, a belief-crutch of sorts that will get Dumbo off the ground by the power of positive thinking. Now imagine a variation on that scene. Imagine that one of the other crows, a village skeptic who is smart enough to see what trick is being played but not smart enough to see its virtue, starts trying to inform Dumbo of the truth as he perches on the cliff edge, feather held tightly. "*Stop that crow!*" the children would shriek. Stifle that smarty-pants, quick, before he ruins it for Dumbo!

In the eyes of some, I am that crow. Look out, they warn. This person is up to some serious mischief, however well intentioned. He insists on talking about topics that are better left unexplored. "Shhh! You'll break the spell." This admonition is not just for fairy tales; it is sometimes quite appropriate in real life. A fact-laden disquisition on the biomechanics of sexual arousal and erection is not a good topic during foreplay, and reflections on the social utility of ceremony and costume are unwelcome in a funeral oration or wedding toast. There are times when we are wise to divert our attention from scientific detail, when ignorance is indeed bliss. Is this another such case?

Dumbo's flying just happens to depend on Dumbo's believing he can fly. This isn't a necessary truth; if Dumbo were a bird (or just a more self-confident elephant!), his talent wouldn't be so fragile, but being who he is, he needs all the moral support he can get, and our scientific curiosity shouldn't be allowed to interfere with his delicate state of mind. Is free will like that? Isn't it at least probable that having free will depends on believing you have free will? And if it is even probable, shouldn't we avoid expressing doctrines that might rightly or wrongly undermine that belief? If we can't go along with the gag, aren't we at least obliged to button our lips or change the topic of conversation? Certainly there are those who think so.

In the many years that I have been working on this problem, I've come to recognize a pattern. My fundamental perspective is *natu-*

ralism, the idea that philosophical investigations are not superior to, or prior to, investigations in the natural sciences, but in partnership with those truth-seeking enterprises, and that the proper job for philosophers here is to clarify and unify the often warring perspectives into a single vision of the universe. That means welcoming the bounty of well-won scientific discoveries and theories as raw material for philosophical theorizing, so that informed, constructive criticism of both science and philosophy is possible. As I present the fruits of my naturalism, my materialist theory of consciousness (e.g., in *Consciousness Explained,* 1991A), and my account of the mindless, purposeless Darwinian algorithms that created the biosphere and all its derivative products—both our brains and our brainchildren—(e.g., in *Darwin's Dangerous Idea,* 1995), I encounter pockets of uneasiness, a prevailing wind of disapproval or anxiety quite distinct from mere skepticism. Usually this discomfort is muffled, like a faint rumble of distant thunder, a matter of wishful thinking almost subliminally distorting the agenda. Often, after the interlocutors have exhausted their supply of objections, someone will expose the hidden agenda that has been driving their skepticism: "That's all very well, but then what about free will? Doesn't your view destroy the prospect for free will?" This is always a welcome response, since it supports my conviction that concern about free will is the driving force behind most of the resistance to materialism generally and neo-Darwinism in particular. Tom Wolfe, who is tuned into the zeitgeist as well as anybody, has captured this motif in a piece with the suitably frantic title "Sorry, but Your Soul Just Died." It is about the rise of what he somewhat confusedly labels "neuroscience," whose chief ideologue he identifies as E. O. Wilson (who is, of course, not a neuroscientist at all, but an entomologist and sociobiologist), along with his henchmen, Richard Dawkins and me. Wolfe thinks he sees the handwriting on the wall:

> Since consciousness and thought are entirely physical products of your brain and nervous system—and since your brain arrived fully imprinted at birth—what makes you think you have free will? Where is it going to come from? (Wolfe 2000, p. 97)

I have an answer. Wolfe is just wrong. For one thing, your brain isn't "fully imprinted at birth," but that's the least of the misunderstandings behind this widespread resistance to naturalism. Naturalism is no enemy of free will; it provides a _positive_ account of free will, one that handles the perplexities better, in fact, than those views that try to protect free will from the clutches of science with an "obscure and panicky metaphysics" (in P. F. Strawson's fine phrase). I presented a version of it in my 1984 book, _Elbow Room: The Varieties of Free Will Worth Wanting_. But I find that people often doubt that I could possibly mean what I say. They are convinced, along with Tom Wolfe, that _of course_ materialism must find no room for free will, and whereas Wolfe is at least sometimes mordantly cheery about this ("I love talking to these people—they express an uncompromising determinism"), others are not. Brian Appleyard, for instance, has written several alarm calls in the form of books, but according to yet another alarmist, Leon Kass, he himself has been seduced:

> Appleyard dislikes, quite properly, the implications of genocentrist thinking and expresses the hope that it may yet be found mistaken; in any case, he insists that it must be resisted. But he is not himself philosophically equipped to show what is wrong with it. Worse, he appears to be an unwitting victim of such thinking, taken in by the inflated pronouncements of the most reductionist and grandiose bioprophets: Francis Crick, Richard Dawkins, Daniel Dennett, James Watson and E. O. Wilson. (Kass 1998, p. 8)

Determinism, genocentrism, reductionism—beware these grandiose bioprophets; they are about to subvert all that is precious! Faced so often with these condemnations (and misrepresentations, as we shall see), I have recognized the need for something in the way of an _apologia_. Am I doing something irresponsible in promulgating these ideas so vigorously?

Scholars in their traditional ivory towers have typically not worried much about their responsibility for the _environmental impact_ of their work. The laws of libel and slander, for instance, exempt none of us, but most of us—including scientists in most fields—do not typically make assertions that, independently of libel and slander considerations, might bring harm to others, even indirectly. A handy measure of this fact is the evident ridiculousness we discover in the idea of malpractice insurance for literary critics, philosophers, mathematicians, historians, cosmologists. What on earth could a mathematician or literary critic do, in the course of executing her professional duties, that might need the security blanket of malpractice insurance? She might inadvertently trip a student in the corridor or drop a book on some-body's head, but aside from such outré side effects, our activities are paradigmatically innocuous. One would think. But in those fields where the stakes are higher—and more direct—there is a long-standing tradition of being especially cautious, and of taking particular respon-sibility for ensuring that no harm results (as explicitly honored in the Hippocratic Oath). Engineers, knowing that the safety of thousands of people may depend on the bridge they design, engage in focused exer-cises with specified constraints posed to determine that, according to all current knowledge, their designs are safe and sound. When we aca-demics aspire to have a greater impact on the "real" (as opposed to "academic") world, we need to adopt the attitudes and habits of these more applied disciplines. We need to hold ourselves responsible for what we say, recognizing that our words, if believed, can have profound effects for good or ill.

Not just that. We need to recognize that our words *might be misunderstood,* and that we are to some degree just as responsible for *likely* misunderstandings of what we say as we are for the "proper" effects of our words. The principle is familiar: The engineer who designs a product that is potentially dangerous if misused is just as responsible for the effects of misuse as for the effects of appropriate use, and must do whatever is necessary to ward off dangerous misapplica-tions of the product by the uninitiated. Saying the truth as best we can muster is our first responsibility, but truth is not enough. The truth can hurt, especially if people misunderstand it, and any academic who thinks that truth is a sufficient defense for any assertion has probably not thought very hard about the possibilities. Sometimes the likelihood of misunderstanding (or other misuse) of one's *true statements,* and the anticipatable harm such misunderstanding could propagate, will be so great that one had better shut up.

A former student of mine, Paulina Essunger, developed a vivid example that takes the issue out of philosophical fantasy-land into cold reality. She has worked in AIDS research, and knows the perils that face that field well, so I will call her example the Peril of Paulina:

> Let's say I were to "discover" that HIV can be eradicated from an infected individual under ideal circumstances (total patient com-pliance, total absence of events inhibiting drug-action such as

nausea, etc., total absence of contamination with extraneous virus strains, and so on) with four years of a certain therapeutic regime. I can be wrong about this. I can be wrong in a quite simple, straightforward way. Say I've miscomputed something, misread some data, misjudged the enrolled patients, or perhaps extrapolated too generously. *I could also be wrong in publishing these results even if they are true, because of their potential environmental impact.* (Further, the media could be wrong in carrying the story, could be wrong in *how* they carry the story. But some of their responsibility seems to fall back on me. Especially if I use the word "eradicate," which in viral contexts usually refers to wiping the virus off the face of the earth, not "merely" ridding one infected individual of it.) For instance, an irrational complacency may spread among, let's say, male homosexuals: "AIDS is curable now so I don't have to worry about it." The incidence of unprotected high-risk sex in this group might rise again due to this complacency. Moreover, the widespread prescription of the treatment might lead to a dramatic spread of resistant virus in the infected population due to periodic patient non-compliance. (Essunger, personal correspondence)

In the worst case, you could have a cure for AIDS, *know* you have a cure for AIDS, and yet be unable to find a way of making that knowledge publicly available in a responsible way. It is no good fuming at the complacency or recklessness of the at-risk community, no good blaming the irresolute patients who abandon their treatments in midcourse—these are predictable and natural (if lamentable) effects of the impact your publication would have. You should explore all the practical avenues for preventing these abuses of your discovery, of course, and make plans to implement whatever safeguards you can, but maybe, in the worst case, the imaginable benefits of your discovery are simply unattainable: You just can't get there from here. This would be not just a serious dilemma; it would be a tragedy. (Her hypothetical case is, of course, already coming true in some regards: Optimism about an impending cure has already led to dangerously relaxed attitudes about safe sexual practices in at-risk groups in the Western world.)

This, then, is a possibility in principle, but is it at all likely that such systematic sources of frustration confront my attempt to promulgate a naturalist "cure" for the free will problem? In fact, there are a few such sources, and they are indeed frustrating. There are various guardians of the public good who—with the best of intentions—want to *stop that crow!* They are prepared to take whatever steps they can to discourage, squelch, or discredit those they see as breaking the spell, before some serious harm is done. They have been at it for many years, and while their campaigns have grown threadbare, and their simple fallacies have been exposed over and over by their scientific colleagues, the debris from their campaigns continues to pollute the atmosphere of the discussions, distorting the understanding of the general public on these topics. For instance, the biologists Richard Lewontin, Leon Kamin, and Steven Rose once said that they consider themselves

> a fire brigade, constantly being called out in the middle of the night to put out the latest conflagration, always responding to immediate emergencies, but never with the leisure to draw up plans for a truly fireproof building. Now it is IQ and race, now criminal genes, now the biological inferiority of women, now the genetic fixity of human nature. All of these deterministic fires need to be doused with the cold water of reason before the entire intellectual neighborhood goes up in flames. (Lewontin et al. 1984, p. 265)

Nobody ever said a fire brigade had to fight fair, and this brigade throws a lot more than the cold water of reason on those they see as incendiary. They are not alone. Coming from the opposite pole of the political spectrum, the religious right has also mastered the art of refutation by caricature, and pounces on every opportunity to replace cautiously expressed articulations of the evolutionary facts with sensationalized oversimplifications that they can then hoot at and warn the world about. I agree with the critics on both left and right that there have been *some* unfortunate overstatements and oversimplifications by some of those they target, and I also agree that such lapses from responsibility *can* have truly pernicious effects. Moreover, I don't challenge their motives or even their tactics; if I encountered people conveying a message I thought was so dangerous that I could not risk giving it a fair hearing, I would be at least strongly tempted to misrepresent it, to caricature it for the public good. I'd want to make up some good epithets, such as *genetic determinist* or *reductionist* or *Dar-*

winian fundamentalist, and then flail those straw men as hard as I could. As the saying goes, it's a dirty job, but somebody's got to do it. Where I think they go wrong is in lumping the responsible, cautious naturalists (like Crick and Watson, E. O. Wilson, Richard Dawkins, Steven Pinker, and myself) in with the few reckless overstaters, and foisting views on us that we have been careful to disavow and to criticize. As a strategy it is clever: If you really think you have to tar something, use a broad brush, just to be safe; don't let the evil guys hide behind a shield of respectable hostages! But it does have the effect of assailing some natural allies with friendly fire, and to be blunt, it is dishonest, however well intentioned.

The Peril of Paulina that we naturalists face is that whenever we put forward circumspect, precise versions of our positions, some of these guardians of the public good turn their cleverness to transforming our careful claims into sound bites that are indeed foolish and irresponsible. I have found that the more care I devote to making my message clear and compelling, for instance, the more suspicious these guardians become. What they say, in paraphrase, is this: "Don't pay attention to all the caveats and complications masked by slick rhetoric! All he's *really* saying is that you don't have consciousness, you don't have a mind, you don't have free will! We're all just zombies and nothing matters—that's what he's *really* saying!" How can I deal with this? (For the record, that's *not* what I'm really saying.) And to make matters worse, there are some serious defections and disagreements within our supposedly monolithic camp of "Darwinian fundamentalists." For instance, Robert Wright, whose recent book *Nonzero: The Logic of Human Destiny* is in most regards a fine exposition of many of the themes I will be presenting here, finds he is unable to endorse the central claim (as I see it) of our position:

> Of course the problem here is with the claim that consciousness is "identical" to physical brain states. The more Dennett et al. try to explain to me what they mean by this, the more convinced I become that what they really mean is that consciousness doesn't exist. (Wright 2000, p. 398)

Wright retreats, alas, to the mystical vision of Teilhard de Chardin after several hundred fine pages of stalwart naturalistic demystification. (A less radical, but more frustrating, defection is Steven Pinker [1997], whose continued dalliance with mysterian doctrines of consciousness is itself a mystery. Nobody's perfect.)

Evidently the stakes are high. What we have here looks like an evolutionary arms race, with escalation on both sides. But note that instead of responding by trying to out-caricature my opponents, I am wheeling out a different weapon on our side: I am trying to plant the seed of suspicion in you that some of these eminent critics of ours may even know in their hearts that we are right. The crow was right, after all, but still, they think, *Stop that crow!* As we shall see in later chapters, some of the most popular objections to a naturalistic account of free will are propelled by fears rather than reasons. The fears themselves are reasonable enough; if you think the box being offered to you might be Pandora's box, by all means put suspicion on a hair-trigger and exhaust all your objections before letting the box be opened, for then it might be too late.

Why, in the face of this heated resistance, do I persist in attempting to present my view, especially since I acknowledge that it is not obvious that it mightn't do some harm? (The critics make the peril greater, of course, by insisting on characterizing the views in dangerous versions; they are playing chicken with us naturalists, in effect.) Because I think it is high time for Dumbo to be weaned from his magic feather. He doesn't need it, and the sooner he learns this, the better. In the movie, you may recall, the feather slips from Dumbo's grasp at a crucial moment, as he is hurtling to his doom, and at the last instant he wises up and saves himself by spreading his ears and pulling out of the dive. It's called growing up, and I think we are ready to grow up. Why is Dumbo better off without his myth of magic? Because he is less dependent, more enabled, more autonomous in the undeluded state. I will try to show that *some* of our traditional ideas about free will are just plain wrong, and moreover that they actually get matters backward, in ways that create serious problems for the future of free will on this planet. For instance, an undeluded view of free will can clarify some of our ideas about punishment and guilt, and allay some of our anxieties about what I call the Specter of Creeping Exculpation (is science going to show us that nobody ever deserves punishment? Or praise, for that matter?). It can help reestablish the proper role of moral education, and even explain the important role religious ideas have played in the past in sustaining morality in society, a role that is no longer being well played by religious ideas but which we discard at our peril. If we persist with the myths, if we dare not turn them in for scientifically sound replacements—which are available—our flying days may be numbered. The truth really will set you free.

Chapter 1

A naturalistic account of how we and our minds evolved seems to threaten the traditional concept of free will, and fear about this prospect has distorted the scientific and philosophical investigation of these issues. Some who have sensed the dangers of these new discoveries about ourselves have seriously misrepresented them. The implications of our newfound knowledge of our origins will prove, on calm examination, to support a stronger, wiser doctrine of freedom than the myths it must replace.

Chapter 2

Our thinking about determinism is often distorted by illusions that can be banished with the help of a toy model, in which simple entities can evolve that are capable of avoiding harm and reproducing themselves. This demonstrates that the traditional link between determinism and inevitability is a mistake, and that the concept of inevitability belongs at the design level, not the physical level.

Notes on Sources and Further Reading

The full reference for books and articles referred to in the text (e.g., Wolfe 2000) can be found in the Bibliography at the end of the book. For each chapter I will provide some further comments and signposts to other sources on the topics discussed.

It may have occurred to some readers that I get off to a bad start in this book by contradicting myself on page 3. First I deny that we have souls in addition to trillions of robotic cells and then I blithely observe that we are conscious: "Since I am conscious and you are conscious, we must have conscious selves that are *somehow* composed of these strange little parts." You may find yourself strongly tempted to agree with Robert Wright that I am actually claiming that consciousness doesn't exist. It would be a shame if you allowed that conviction to distort your reading of the rest of the book, so please try to reserve judgment, on the off chance that Wright is wrong! My uncompromising materialism really is an integral feature of the view I will be defending, and I wanted to be up front about it, even at the risk of creating antagonism and skepticism in those who still hanker for a dualistic account of consciousness. The articulation and defense of this material theory of consciousness can be found in my books mentioned above, and is further elaborated and defended against various recent criticisms in my Jean Nicod Lectures, delivered in Paris in November 2001 (forthcoming D), as well as in a series of papers published or forthcoming in a variety of journals and volumes and also available on my Web site: http://ase.tufts.edu/cogstud.

The philosophical literature on free will is enormous, and only a small fraction of the recent work on the topic will receive attention in these pages. Those that are discussed will provide plenty of threads leading to the rest. Two outstanding books by non-philosophers have been published in the year I was putting the final touches on my book, and these should be read by anybody interested in the topic: George Ainslie's *Breakdown of Will* (2001) and Daniel Wegner's *The Illusion of Conscious Will* (2002). I have worked brief reflections on these two books into my own, but the richness of their contributions goes well beyond what can be surmised from those reflections.

Chapter 2

A TOOL FOR THINKING ABOUT DETERMINISM

Determinism is the thesis that "there is at any instant exactly one phys-ically possible future" (Van Inwagen 1983, p. 3). This is not a particu-larly difficult idea, one would think, but it's amazing how often even very thoughtful writers get it flat wrong. First, many thinkers assume that determinism implies inevitability. It doesn't. Second, many think it is obvious that *in*determinism—the denial of determinism—would give us agents some freedom, some maneuverability, some elbow room, that we just couldn't have in a deterministic universe. It wouldn't. Third, it is commonly supposed that in a deterministic world, there are no *real* options, only apparent options. This is false. *Really?* I have just contradicted three themes so central to discussions of free will, and so seldom challenged, that many readers must suppose I am kidding, or using these words in some esoteric sense. No, I am claiming that the complacency with which these theses are commonly granted without argument is itself a large mistake.

Some Useful Oversimplifications

These errors lie at the heart of the misconceptions about free will and freedom more generally, so before we can make any progress on under-standing how freedom could evolve (in a universe that may well be deterministic), we need to equip ourselves with some corrective devices, some tools for thinking that will make us less vulnerable to the siren songs of these powerful illusions. (If you have an aversion to philosophical argumentation about determinism, causation, possibility, necessity, and the indeterminism of quantum physics, you may skip ahead to Chapter 5, but you must then forswear all reliance on these three "obvious" propositions, no matter how intuitive they strike you, and take it on faith when I assure you that they are the false friends of a thousand misguided discussions. I almost guarantee that you cannot keep that resolution, however, so a better choice is to plunge into my demonstrations of these errors, which have their rewards and surprises, and presuppose no background expertise.)

In Thomas Pynchon's novel *Gravity's Rainbow,* a character makes the following portentous speech:

> But you had taken on a greater, and more harmful, illusion. The illusion of control. That A could do B. But that was false. Completely. No one can *do.* Things only happen. (Pynchon 1973, p. 34)

Pynchon's speaker has concluded that since atoms can't *do* anything, and people are made of atoms, people can't *do* anything either, not really. He is right that there is a difference between doing and mere happening, and he is right that there is a harmful illusion lurking in our attempts to understand this difference, but he gets the illusion backward. It is not the mistake of treating people as if they weren't composed of lots of *happening* atoms (they are), but *almost* the reverse: treating atoms as if they were little people *doing* things (they aren't). It arises when we overextend the categories appropriate to evolved *agents* onto the wider world of physics. The world of *action* is the world we live in, and when we try to impose the perspective of that world back down onto the world of "inanimate" physics, we create a deeply misleading problem for ourselves.

Getting clear about this aspect of the complex relationship between fundamental physics and biology sounds terrifying, but fortunately, there is a *toy* version of that relationship that is just what we need. The difference between a toy and a tool can evaporate if the toy can help us understand things that are otherwise too complex for us to keep track of. Science often uses toy models to great advantage. Nobody has seen an atom, but we all know what an atom "looks like": a tiny solar system, with a nucleus like a tight bunch of grapes surrounded by electrons orbiting every which way in their little halos. This familiar friend, the Bohr model (Figure 2.1), is of course hugely over-simplified and distorted, but for many purposes it's a great way to think about the basic structure of matter.

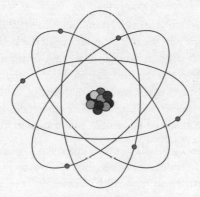

Figure 2.1 Bohr Atom

Becoming just as familiar in our common imagination is the gigantic Tinkertoy construction of a double helix with lots of rungs, the Crick-Watson model of the DNA molecule (Figure 2.2). It, too, is a useful oversimplification.

Figure 2.2 DNA Double Helix

The French physicist and mathematician Pierre-Simon Laplace gave us a usefully simple and vivid image of determinism almost two centuries ago, and it has structured our imaginations, and hence our theories and debates, ever since.

> An intellect which at any given moment knew all the forces that animate Nature and the mutual positions of the beings that comprise it, if this intellect were vast enough to submit its data to analysis, could condense into a single formula the movement of the greatest bodies of the universe and that of the lightest atom: for such an intellect nothing could be uncertain; and the future just like the past would be present before its eyes. (Laplace 1814)

Give this all-knowing intellect, often known as *Laplace's demon,* a complete snapshot of "the state of the universe," showing the exact location (and trajectory and mass and velocity) of every particle at that instant, and the demon, using the laws of physics, will be able to plot every collision, every rebound, every near miss in the next instant, updating the snapshot to yield a new state description of the universe, and so on, for eternity.

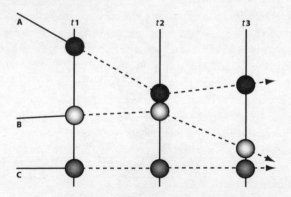

Figure 2.3   Laplacean Snapshot

In Figure 2.3, this snapshot zooms in at time *t1* on just three of the atoms in the world, on their various trajectories, and the demon uses this information to predict the collision and rebound of two of them at *t2,* leading to the positions at *t3* and so on. A universe is *determinis-*

tic if there are transition rules (the laws of physics) that *determine exactly* which state description follows any particular state description. If there is any slack or uncertainty, the universe is indeterministic.

There are too many fudge factors in this simple vision as it stands: How exact must a state description be? Must we plot every subatomic particle, and just which properties of the particles need to be included in the description? We can anchor these slippery factors arbitrarily by adopting another simplifying idea, W.V.O. Quine's (1969) proposal that we restrict our attention to simple imaginary universes, which he calls "Democritean" universes, in honor of Democritus, the most inventive of the ancient Greek atomists. A Democritean universe consists of some "atoms" moving about in "space." That's all. The atoms in a Democritean universe are not modern atoms full of quantum complexities but truly *a-tomic* (unsplittable, unsliceable) atoms, tiny uniform points of matter with no parts at all, rather like those postulated by Democritus. The space they inhabit must be made ultrasimple, too, by *digitizing* it. Your computer screen is a good example of a digitized *plane,* a two-dimensional array of hundreds of rows and columns of tiny *pixels,* little squares, each of which has, at each instant, one of a finite set of different colors. We want to digitize a space, a three-dimensional volume, so we need cubes—*voxels,* in the language of computer graphics. Imagine a universe composed of an infinite latticework of tiny cubical voxels, each one either utterly empty or utterly full (containing exactly one atom). Each voxel has a unique location or address in the latticework, given by its three spatial coordinates, $\{x, y, z\}$. Just as every computer color graphics system has a certain range of values—different shades of color—that each pixel can take on, in a Democritean universe, every voxel that isn't empty (value 0) contains one of a limited number of different types of atoms. It may help to think of them as different colors, such as gold, silver, black (carbon), yellow (sulfur). Just as we can define the set of all possible computer-screen images (for any particular pixel-color system) as the set of all permutations of fillings of the pixels with the defined colors, we can define the set of all Democritean-universe moments as the set of all permutations of fillings of all the voxels in space with the various sorts of atoms.

Now when we want to confront Laplace's demon with a "complete" snapshot from which to work, we can say exactly what we need to provide: a *state description* of a *Democritean universe,* which lists the values of every voxel at some instant. So part of state description S_k might read:

> at time t:
>> voxel $\{2,6,7\}$ = silver,
>> voxel $\{2,6,8\}$ = gold,
>> voxel $\{2,6,9\}$ = 0,
>> . . . and so forth.

We don't have to worry about how "fine-grained" to make our description, since a Democritean universe has a defined limit, a smallest difference, and we can compare any two state descriptions of the universe and discover any corresponding voxels that are differently occupied. As long as there are a finite number of different elements (gold, silver, carbon, sulfur . . .) we can put all the state descriptions in order—alphabetical order, in effect—by voxel and the element occupying it. State description 1 is the empty universe at time t; state description 2 is just like 1 except for having a single aluminum atom occupying voxel $\{0,0,0\}$; state description 3 moves that lone aluminum atom to voxel $\{0,0,1\}$; and so forth, all the way to the last state description (in alphabetical order), in which the universe is filled—every voxel—with zinc! Now add time, the fourth dimension. Suppose that at the next "instant," the gold atom at $\{2,6,8\}$ in S_k moves east one voxel. Then in S_{k+1},

> at time $t + 1$:
>> . . .
>> voxel $\{3,6,8\}$ = gold.

Think of each "instant" of time as like a frame of computer animation, specifying the color or value of each voxel at that instant. This digitizing of space and time permits us to count differences and similarities, and to say when two universes, or regions or periods of universes, are exactly alike. A series of state descriptions, one for each successive "instant," yields the history of a whole Democritean universe, for however long that universe lasts—from its Big Bang to its Heat Death (or whatever replaces these openings and closings in these imaginary worlds). *In other words, a Democritean universe is like a 3-D digital video of some length or other.* We can cut time as fine as we like; thirty frames a second (like a movie) or thirty trillion frames a second, depending on our purposes. The size of the voxels is minimal: one indivisible atom per voxel, max. Quine proposed a further simplification: Imagine that the atoms are all alike (rather like electrons), so we can treat each voxel as either empty (value $= 0$) or full (value $= 1$). This option is just like replacing a color screen with a black-and-white screen, a simplification good for some purposes, as we shall see, but not necessary.

How many different ways are there of filling voxels with colors (or just with 0 and 1)? Even when we keep the size of a universe not just finite but tiny, the number of possibilities gets huge in a hurry. A universe consisting of just eight voxels (making a two-by-two cube) and one kind of atom (empty or full, 0 or 1), and lasting only 3 "instants," has already more than 16 million different variations ($2^8 = 256$ different state descriptions, which can be put together in 256^3 different series of three). A second's-worth of the universe contained in a single sugar cube (at the *slow* rate of 30 frames a second and taking the cube to be *only* a million atoms wide) would be a number of states beyond imagining.

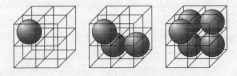

Figure 2.4   Three of the 256 different states of an
8-voxel Democritean universe.

In *Darwin's Dangerous Idea,* I introduced the term "Vast" as a name for numbers that, though finite, are Very much larger than ASTronomical quantities. I used it to characterize the not-really-infinite number of books in Jorge Luis Borges's imaginary Library of Babel, the set of all possible books, and by extension, the number of possible genomes in the Library of Mendel, the set of all possible genomes. I also coined a reciprocal term, "Vanishing," to characterize, for instance, the subset of *readable* books, nearly invisible within the Library of Babel. Let's call the set of all possible Democritean universes, all the logically possible combinations of atoms in space and time, the Library of Democritus. The Library of Democritus is mind-bogglingly large, no matter how tightly we restrict it to a particular finite set of parameters (types of atoms, durations, etc.). Things get interesting when we look at particular subsets of the Library. Some universes in the Library of Democritus are practically empty, and others are full of stuff; some have lots of change over time and others are static—the same state description, repeated forever. In some the change is utterly random—one instant of atomic confetti after another, with individual atoms flicking in and out of existence—and others show patterns of regularity and hence predictability. Why do some universes show patterns? Just because the Library of Democritus contains all the logically possible universes, so *every possible pattern* whatsoever is to be found somewhere in it; the only rule is that each state description should be complete and self-consistent (only one atom to a voxel).

Once we start imposing additional rules about what can be adjacent to what, and about how different state descriptions should succeed each other in time, we can get to more interesting subsets of the Library. For instance, we could prohibit the "annihilation of matter" by a rule that says that every atom that exists at time *t* has to exist somewhere at time *t* + 1, though it can move to a new voxel if that voxel is unoccupied. This guarantees that the universe never loses an atom as time passes. (More precisely, we "prohibit" this by just ignoring the Vastly many universes that don't obey this rule and restricting our attention to the Vast but Vanishing subset of those that do obey it: "Consider the set S of universes in which the following rule always holds. . . .") We could set up a speed limit (rather like the speed of light) by adding that an atom can move only to a neighboring voxel in the next instant, or we could permit longer leaps. We could say that matter *can* be annihilated—or created—under such-and-such conditions: For instance, we could have the rule that whenever two gold atoms are stacked one on top of the other, in the next instant they disappear, and in the lower voxel an atom of silver comes into existence. Such transition rules are tantamount to the fundamental laws of physics that hold in each imaginary universe, and we can usefully look at sets of universes in which these regularities are the same, whatever other differences there might be. Suppose, for instance, that we want to "hold physics constant" but vary the "initial conditions"—the state of the universe at its debut instant. We then consider the set of universes in which a particular transition rule or set of rules always holds but the starting-state descriptions are as varied as we like. This is rather like restricting our attention, in the Library of Babel, to those books written in (grammatical) English; there are regularities in the transition from character to character ("*i*" *before* "*e*" *except after* "*c*" . . . and *Every question begins with a capital letter and ends with a question mark.* . . .), but the topics covered are as varied as can be.

A better analogy between Borges's Library of Babel and our Library of Democritus would be the existence, in the Library of Babel, of Vastly many books that start out just fine—as novels or histories or chemistry textbooks—but then suddenly degenerate into nonsensical word salad, typographical gibberish. For every book that can be read cover to cover for enjoyment and profit, there are Vast numbers of volumes that start out well, with the regularities of grammar, vocabulary, story line, character development, and so forth that are prerequisite for *making sense,* but then degenerate into patternlessness. There is no *logical* guarantee that a book that starts well will continue well. The same is true of the Library of Democritus. This was David Hume's point, back in the eighteenth century, when he observed that even though the sun has risen every day so far, *there is no contradiction* in the supposition that tomorrow will be different, that the sun will not rise. To translate his observation into Library of Democritus talk, note that there is a set of universes, A, in which the sun *always* rises, and there is a set of universes, B, in which the sun always rises *until* [*say*] *September 17, 2004, at which point something else happens.* There's nothing contradictory about those worlds—they just don't turn out to "obey" the physics that always holds in universes in set A. Hume's point can be put this way: No matter how many facts you gather about the past of the universe you find yourself in, you can never prove, logically, that you're in a universe in set A, since for each universe in set A, there are Vastly many universes in set B that are identical to it at every voxel/time up to September 17, 2004, and then diverge in all manner of surprising or fatal directions!

As Hume noted, we *expect* the physics that has held so far in our world to hold in the future, but we cannot prove by pure logic that it will oblige us. We've had conspicuous success discovering regularities that have held in the past in our universe, and we've even learned how to make real-time predictions, about seasons and tides and falling objects and what you'll find if you dig here, or dissect there, or heat this or mix that with water, and so forth. These transitions are so regular, so unexceptioned in our experience, that we have been able to codify them and project them imaginatively into the future. So far so good; it has worked like a charm, but there are no logical guarantees it will continue to work. Still, we have some reason to believe that we inhabit a universe in which this process of discovery can go on *more or less* indefinitely, yielding ever more specific, reliable, detailed, accurate predictions based on the regularities we have observed. In other words, we may take ourselves to be finite, imperfect approximations of Laplace's demon, but we can't prove, logically, that our success will continue, without presupposing the very regularities whose universality and eternity we would like to establish. And there are some reasons, as we shall see, to conclude that there are absolute limits on our capacity to predict the future. Whether these limits have any implications about our self-image as agents making "free" decisions and choices, for which we might properly be held responsible, is one of the treacherous questions we need to address, and we are approaching it gingerly, getting clear about simpler issues first. We're gradually approaching our target, *determinism,* by closing in on a Vast but Vanishing neighborhood in the still Vaster space of logically possible universes.

Some sets of Democritean universes have transition rules that are deterministic, and some don't. Consider the set of universes in which we specify that whenever an atom is surrounded by empty voxels it has a one-in-thirty-six chance of simply vanishing—otherwise, it stays put in the next instant. In such universes it is as if Nature rolled some dice whenever such an atom got itself isolated in this way; if the dice come up snake eyes, the atom "dies"; otherwise, it lives another instant and Nature rolls the dice again, unless that atom has just acquired a neighbor. This would be an *indeterministic* physics, which does not specify what happens next in all regards but leaves some of the transitions to mere probability. Laplace's demon would have to wait to see how the dice came up before continuing to predict the future. Other sets of universes obey transition rules that leave nothing to chance, that specify exactly what voxels are occupied by what atoms in the next moment. These are the deterministic universes. There are, of course, kazillions of different ways the transition rules for Democritean universes could be deterministic or indeterministic.

How do we *tell* what transition rules govern a particular Democritean universe? We can *stipulate* a rule and then consider what we must or might find to be true in all possible members of the set obeying the rule, but if we are somehow given a particular Democritean universe to study, the only thing we can do is examine the entire history of all its voxels and see what regularities—if any—hold. We can break the job into natural parts by looking for regularities that hold in the early going and seeing if they continue to hold all the way forward. Bearing in mind Hume's ominous discovery that we can never prove that the future will be like the past, we can nevertheless set out to find what regularities we can and make the huge but tempting wager—what do we have to lose?—that the future *will* be like the past, that we are not in one of those bizarre universes that leads us down the garden path only to disappoint us by going haywire after a longish period of regularity.

We now have a way of sorting Democritean universes into the deterministic, the indeterministic, and then all the junk—we might call these the *nihilistic* universes in which there is no permanent regularity of transition at all. Notice that on this construal, *all there is* to being deterministic or indeterministic is always exhibiting one sort of regularity or another—either a regularity with incliminable probabilities less than one, or a regularity in which all such probability is absent. There is no room, in other words, for the claim that two Democritean universes are exactly alike at each voxel/time, but one of them is deterministic and the other is indeterministic.[1]

The difference between deterministic and indeterministic Democritean universes is now clear, but the best way of understanding just what it means (and what it *doesn't* mean!) is to pamper our overwhelmed imaginations even more and consider a still simpler toy image of determinism. First, let's drop from three dimensions to two (from voxels down to pixels), and let's also avail ourselves of Quine's black-and-white-only option, so that each pixel is either ON or OFF at

---

1. Indeed, by definition, no *two* Democritean universes are exactly alike at each voxel/time. One of the virtues of Quine's simplification is that it lets us count universes the same way we count *editions* of books: If all the same elements are in the same places at the same times, that establishes *identity*. Quine's proposed taming of possible worlds also eschews the dubious idea that we need to know the *identity* of the individual atoms—not just their type, carbon or gold—to identify voxel contents from one universe to another. (Maven alert: This is not standard possible worlds lore; it avoids familiar problems of transworld identity.)

any instant. We have now landed on the plane where Conway's Game of Life spins out its astonishing patterns. This audaciously oversimplified toy model of determinism was developed in the 1960s by the British mathematician John Horton Conway. Conway's Life vividly illustrates just the ideas we need in a way that requires no technical knowledge of either biology or physics, and no math beyond the simplest arithmetic.

From Physics to Design in Conway's Life World

The complexity of a living individual minus its ability to anticipate (in respect of its environment) equals the uncertainty of the environment minus its sensibility (in respect of that particular living individual).

—Jorge Wagensberg, "Complexity versus Uncertainty"

Consider, then, a two-dimensional grid of pixels, each of which can be ON or OFF (full or empty, black or white).[2] Each pixel has eight neighbors: the four adjacent cells: north, south, east, and west, and the four diagonals: northeast, southeast, southwest, and northwest. The state of the world changes between each tick of the clock according to the following rule:

> *Life Physics:* For each cell in the grid, count how many of its eight neighbors is ON at the present instant. If the answer is exactly two, the cell stays in its present state (ON or OFF) in the next instant. If the answer is exactly three, the cell is ON in the next instant whatever its current state. Under all other conditions the cell is OFF.

That's all. This one simple transition rule expresses the entire physics of the Life world. You may find it a useful mnemonic crutch to think of this curious physics in biological terms: Think of cells going ON as births, cells going OFF as deaths, and succeeding instants as generations. Either overcrowding (more than three inhabited neighbors) or isolation (less than two inhabited neighbors) leads to death. But remember, this is just a crutch for the imagination: the two-three rule

---

2. This introduction to Life is drawn, with revisions, from Dennett 1991A and Dennett 1995.

is the basic *physics* of the Life world. Consider how a few simple start-ing configurations play themselves out.

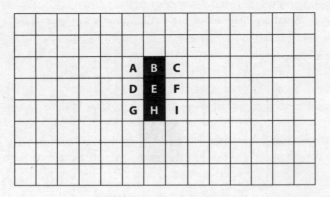

Figure 2.5 Vertical Flasher

Calculate birth cells first. In the configuration shown in Figure 2.5, only cells *d* and *f* have exactly three neighbors ON (dark cells), so they will be the only birth cells in the next generation. Cells *b* and *h* each have only one neighbor ON, so they die in the next generation. Cell *e* has two neighbors ON, so it stays on. So the next instant will look like this:

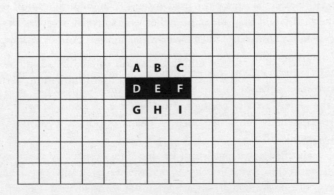

Figure 2.6 Horizontal Flasher

Obviously, the configuration shown in Figure 2.6 will revert back in the next instant, and this little pattern will flip-flop back and forth indefinitely, unless some new ON cells are brought into the picture somehow. It is called a *flasher* or traffic light.

What will happen to the configuration in Figure 2.7?

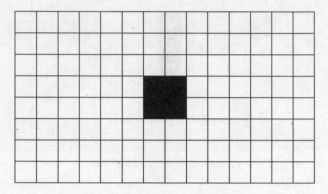

Figure 2.7   Square Still Life

Nothing. Each ON cell has three neighbors ON, so it is reborn just as it is. No OFF cell has three neighbors ON, so no other births happen. This configuration is called a *still life;* there are many different still life configurations that do not change at all over time.

By the scrupulous application of our single law, one can predict with perfect accuracy the next instant of any configuration of ON and OFF cells, and the instant after that, and so forth, so *each Life world is a deterministic two-dimensional Democritean universe.* And to first appearances, it fits our stereotype of determinism perfectly: mechanical, repetitive, ON, OFF, ON, OFF for eternity, with never a surprise, never an opportunity, never an innovation. If you "rewind the tape" and play out the sequel to any configuration again and again, it will always come out exactly the same. Boring! Thank goodness we don't live in a universe like that!

But first appearances can be deceiving, especially when you're standing too close to the novelty. When we step back and consider larger patterns of Life configurations, we are in for some surprises. The flasher has a two-generation period that continues *ad infinitum,* unless some other configuration encroaches. *Encroachment is what makes Life interesting.* Among the periodic configurations are some that swim, amoeba-like, across the plane. The simplest is the *glider,* the five-pixel configuration (Figure 2.8) shown here taking a single stroke to the southeast:

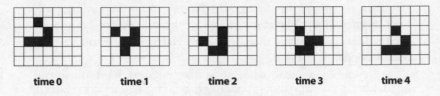

Figure 2.8   Glider

Then there are the eaters, the puffer trains, and space rakes, and a host of other aptly named denizens of the Life world that emerge as recognizable objects at a new level. In one sense, this new level is simply a bird's-eye view of the basic level, looking at large clumps of pixels instead of individual pixels. But, wonderful to say, when we ascend to this level, we arrive at an instance of what I have called the *design level;* it has its own language, a transparent foreshortening of the tedious descriptions one could give at the *physical level.* For instance:

> An eater can eat a glider in four generations. Whatever is being consumed, the basic process is the same. A bridge forms between the eater and its prey. In the next generation, the bridge region dies from overpopulation, taking a bite out of both eater and prey. The eater then repairs itself. The prey usually cannot. If the remainder of the prey dies out as with the glider, the prey is consumed. (Poundstone 1985, p. 38)

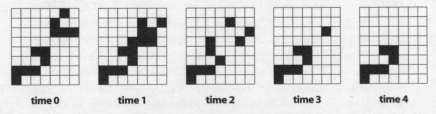

Figure 2.9   Eater Eating a Glider

Notice that something curious happens to our "ontology"—our catalog of what exists—as we move between levels. At the physical level there is no motion, only ON and OFF, and the only individual things that exist, pixels, are defined by their fixed spatial location, $\{x, y\}$. At the design level we suddenly have the motion of persisting objects; it is one and the same glider (though composed each generation of different pixels) that has moved southeast in Figure 2.8, changing shape as it moves; and there is one less glider in the world after the eater has eaten it in Figure 2.9.

Notice too that whereas at the physical level, there are absolutely no exceptions to the general law, at the design level our generalizations have to be hedged: They require "usually" clauses ("the prey usually cannot" repair itself) or "provided nothing encroaches" clauses. Stray bits of debris from earlier events can "break" or "kill" one of the objects in the ontology at this level. Their *salience as real things* is considerable, but not guaranteed. An element of mortality has been introduced. Whereas the individual atoms—the pixels—flash in and out of existence, ON and OFF, without any possibility of accumulating any changes, any history that could affect their later history, larger constructions can suffer damage, a revision of structure, a loss or gain of material that can make a difference in the future. Larger constructions might also happen to be improved, made *less* vulnerable to later dissolution, by something that happened to them. This historicity is the key. The existence in the Life world of structures that can grow, shrink, twist, break, move . . . and in general *persist over time* opens the floodgates to design opportunities.

Rushing in to explore those opportunities is a worldwide fraternity of Life hackers, hobbyists who delight in testing their ingenuity by devising ever more elaborate arrangements on the Life plane that do interesting things. (If you want to explore the Life world, you can download free a fine, user-friendly implementation Life 32 at the Web site http://psoup.math.wisc.edu/Life32.html. It has a library of interesting configurations, and links to other sites. I require my students to explore the Life world, because I have learned that it renders vivid and robust a set of intuitions that are otherwise absent, and helps them think about these issues. In fact—wonder of wonders—it sometimes leads them to *change their minds* about their philosophical positions. So be careful; it can be addictive fun—and it may lead you to abandon your life-defining hatred of determinism!) To become a Life hacker, you simply ascend to the design level, adopt its ontology, and proceed to predict—sketchily and riskily—the behavior of larger configurations or systems of configurations, *without bothering to compute the physical level*. You can set yourself the task of designing some interesting supersystem out of the "parts" that the design level makes available. It takes only a few minutes to get the hang of it, and who knows what you will be able to concoct. What would you get if you lined up a bunch of still life eaters, and then sprayed them with gliders, for instance? After you've dreamed up your design, you can readily test it; Life 32 will swiftly inform you of any overlooked problems in your design stance predictions. You can get a glimpse of the richness of this design level from a few quotes I once pulled off an excellent Life Web site, http://www.cs.jhu.edu/~callahan/lifepage.html#newresults. The Web site is now defunct, sad to say, and don't bother trying to figure out these comments; they are just meant to illustrate the way Life hackers think and talk.

> The loaf reacts with all the junk the R pentomino produces as it naturally transforms into a Herschel, and miraculously reappears some time later leaving no debris at all. It is necessary to prevent the first Herschel glider from hitting the fading remnants of the reaction, and there is no room for an ordinary eater. But luckily a tub with tail and a block can be used instead.

> Dave Buckingham found a faster stable reflector that does not use Paul Callahan's special reaction. Instead, the incoming glider hits a boat to make a B-heptomino, which is converted into a Herschel and moved round to restore the boat. A compact form of the 119-step Herschel conduit is needed here, as is a non-standard still life to cope with the 64 64 77 conduit sequence.

These Life hackers are playing God in their simplified two-dimensional universe, trying to design ever more amazing patterns that will propagate themselves, transform themselves, protect themselves, move themselves around on the Life plane—in short, *do things* in the world, instead of merely flashing back and forth or, worse, just persisting unchanged for eternity (unless something encroaches). As the quotations reveal, the problem that confronts anyone who plays God in this world is that no matter how nice your initial pattern is, it always runs the risk of annihilation, of turning into debris, of being eaten by an eater, of vanishing without a trace.

If you want your creations to persist, they have to be protected. Keeping the physics constant (not changing the basic rule of the Life), the only thing you can play with is the initial state description, but you have so many to choose from! A set of Life worlds only 1 million pixels by 1 million pixels already gives you 2 to the trillionth power of different possible universes to explore—the Library of Conway, a Vast but Vanishing branch of the much, much Vaster Library of Democritus. Some of these Life worlds are very, very interesting, but finding them is harder than hunting for a needle in a haystack. The only way to do it, since random search is practically hopeless, is to think of the search as a design problem: How can I *construct* a Life-form that will *do x* or *do y* or *do z*? And once I've designed something that can *do x,* how can I protect my fine *x-er* from harm once I've constructed it? After all, a lot of precious R&D (research and development) went into designing my *x-er.* It would be a shame if it got smashed before it could do its thing.

How can you make things that will last in the sometimes toxic world of Life? This is an objective, non-anthropomorphic problem. The underlying physics is the same for all Life configurations, but some of them, in virtue of nothing but their *shape,* have *powers* that other configurations lack. This is the fundamental fact of the design level. Let the configurations be as un-human, as un-cognitive, as un-agent-like as you can muster. If they last, what is it about them that explains this? A still life is fine until it gets plowed into. Then what happens? Can it restore itself somehow? Something that can nimbly move out of the way might be better, but how can it get any advance warning of incoming missiles? Something that can eat the incoming debris and profit from it might be better yet. But the rule is: Anything that works is fine. Under that rule, what emerges is sometimes strikingly agent-like, but this may be more a function of a bias in our imagination—like seeing animals in the clouds just because we have lots of animal "templates" in our visual memory—than because it is necessary. In any case, we know a set of tricks that work: a set of tricks that is strongly reminiscent of our own biology. The physicist Jorge Wagensberg has recently argued that this resemblance to life as we know it is no accident. In an essay that does not mention Conway Life, he develops definitions of information, uncertainty, and complexity from which he can derive measures of "independence with respect to the uncertainty of the environment" and use these to show that *persistence,* or what he calls "keeping an identity," in a complex environment depends (probabilistically) on various ways of maintaining "independence"—and these ways include such "passive" measures as "simplification" (like seeds and spores), hibernation, isolation (behind shields and shelters), and sheer size, and above all, the "active" measures that require anticipation. "A biota progresses in a particular environment if the new state of the biota is more independent in respect of the uncertainty of that environment" (Wagensberg 2000, p. 504).

A wall is sometimes a good bargain, if it is strong enough so that nothing can smash it. (Nothing? Well, nothing smaller than G, the most gigantic projectile we've thrown at it yet.) A wall just sits there and takes a beating, not *doing* anything. A mobile protector, on the other hand, must either move in a fixed trajectory, like a sentry marching around the perimeter of a camp; or in a random trajectory, like the swimming-pool vacuum-sweepers that prowl at random, cleaning the walls; or in a guided trajectory that depends on its obtaining some information about the environment through which it moves. A wall that can repair itself is another interesting possibility, but much harder to design than a static wall. These fancier designs, the designs that can take steps to improve their chances, can get quite expensive, since they depend on reacting to information about their circumstances. Their *immediate* surroundings (the eight neighbors around each pixel) are more than informative—they are utterly determining; it is "too late to do anything" about a collision that has begun. If you want your creation to be able to *avoid* some impending harm, it is going to have to be designed either to do the right thing "automatically" (the thing it always does) or to have some way of anticipating it, so that it can be (designed to be) guided by some harbinger or other down a better path.

This is the birth of avoidance; this is the birth of prevention, protection, guidance, enhancement, and all the other fancier, more expensive sorts of *action.* And right at the moment of birth, we can discern the key distinction we will need later on: Some kinds of harms can, in principle, be avoided, and some kinds of harms are unavoidable, or *inevitable,* as we say. Advance warning is the key to avoidance, and this is strictly limited in the Life world by the "speed of light," which is (for all practical purposes) the speed at which simple gliders can swim diagonally across the plane. Gliders, in other words, could be the *photons,* the light particles, in the set of Life universes, and *reacting-to-a-glider* could be a way of turning a mere collision or encroachment into an *informing,* a simplest case of noticing or discriminating. We can see why it is that calamities that arrive at the speed of light must "blindside" any creations they encounter; they are truly inevitable. Slower-moving problems can, in principle, be predicted by any Life-form that can extract guidance from the incoming rain of gliders (or other, slower sources of information) and adjust itself appropriately. It may pick up information about what to expect from other things it encounters, but only if *there is* information in those patterns that is predictive of patterns elsewhere, or at other times. In a totally chaotic, unpredictable environment, there is no hope of avoidance except sheer blind luck.

Notice that I have been intermixing two distinct information-gathering processes in this discussion, which now need to be more clearly separated. First, there is the activity of our hacker Gods, who are free to cast their eyes and minds over huge manifolds of possible Life worlds, trying to figure out what will tend to work, what will be robust and what will be fragile. For the time being, we are supposing that they are truly God-like in their "miraculous" interactions with the Life world—they are not bound by the slow speed of glider-light; they can intervene, reaching in and tweaking the design of a creation whenever they like, stopping the Life world in mid-collision, undoing the harm and going back to the drawing board to create a new design. Wherever *they* can foresee a source of difficulty they can set themselves the task of designing a way of countering it. Their creations will be the unwitting, foresightless beneficiaries of the foresight of the hacker Gods, who have designed them to thrive in just such circumstances. Hacker Gods have their limitations, however, and will economize whenever they can. For instance, they might interest themselves in such questions as: What is the *smallest* Life-form that can protect itself from harm x or harm y, under conditions z (but not under conditions w)? After all, gathering information and putting it to use is a costly, time-consuming process, even for a hacker God. The second possibility is the prospect of the hacker Gods designing configurations that *do their own* information gathering, locally, bound by the physics of the world they inhabit. Expect that any finite creation that uses information will be thrifty, keeping only what it (probably) needs or (probably) can use, given the

vicissitudes in its neighborhood. After all, the hacker God who designs it wants to make it robust enough to fend for itself not in all possible Life worlds but only in any of the set of Life worlds it has some probability of encountering. Such a creation will, at best, be in a position to *act as if it knew* it was living in a particular *sort* of neighborhood, fending off a particular *sort* of harm or securing a particular *sort* of benefit, instead of acting as if it knew exactly which Life universe it inhabited.

Speaking of these smallest avoiders as if they "knew" anything at all involves a large dose of poetic license, since they would be about as close to clueless as you can imagine—they are much simpler than a real-world bacterium, for instance—but it is still a useful way of keeping track of the design work that has gone into them, giving them capabilities to *do things* that any randomly assembled clumps of pixels of about the same size would lack. (Of course, "in principle"—as philosophers love to say—a Cosmic Accident could produce exactly the same constellation of pixels with exactly the same capabilities, but this is an utterly negligible possibility, beyond improbability. Only expensively designed things can do things in the interesting sense.)

Enriching the design stance by speaking of configurations as if they "know" or "believe" something, and "want" to accomplish some end or other is moving up from the simple *design stance* to what I call the *intentional stance.* Our simplest doers have been reconceptualized as *rational agents* or *intentional systems,* and this permits us to think about them at a still higher level of abstraction, ignoring the details of just how they manage to store the information they "believe" and how they manage to "figure out" what to do, based on what they "believe" and "want." We just assume that however they do it, they do it rationally—they draw the right conclusions about what to do next from the information they have, given what they want. It makes life blessedly easier for the high-level designer, just the way it makes life easier for us all to conceptualize our friends and neighbors (and enemies) as intentional systems. *think they are rational*

We can move back and forth between the hacker God perspective and the "perspective" of the hacker God's creations. Hacker Gods have their reasons, good or bad, for designing their creations the way they did. The creations themselves can be clueless about these reasons, but they *are* the reasons those features exist, and if the creations persist, it will be thanks to those features. If, beyond that, the creations

situation - action machine
vs
choice machine

have been designed to gather information to use in action guidance, the situation becomes more complicated. The simplest possibility is that a hacker God has designed a repertoire of reaction-tricks that tend to work well in the environments encountered, analogous to the IRMs (Innate Releasing Mechanisms) and FAPs (Fixed Action Patterns) that ethologists have identified in many animals. Gary Drescher (1991) calls this architecture a *situation-action machine* and contrasts it with the more expensive, more complex *choice machine,* in which the individual creation generates its *own* reasons for doing *x* or *y,* by anticipating probable outcomes of various candidate actions and evaluating them in terms of the goals it also represents (since these goals can change over time, in response to new information gathered). If we ask "at what point" the designer's reasons become the designed agent's reasons, we may find that there is a seamless blend of intermediate steps, with more and more of the design work off-loaded from designer to designed agent. One of the beauties of the intentional stance is that it allows us to see clearly this shift in the distribution of "cognitive labor" between the originating design process and the efforts of the thing designed.

All this fanciful talk about configurations of Life-pixels as rational agents may strike you as outrageous overstatement, a blatant attempt by me to pull the wool over your eyes. It's time for a sanity check: Just how much, in principle, can a designed constellation of Life-pixels *do,* given glider-discrimination and its kin as the "molecules" of the design level, the fundamental building blocks of higher-level Life-forms? This is the question that inspired Conway to create the Game of Life in the first place, and the answer he and his students came up with is staggering. They were able to prove that there are Life worlds—they sketched one of them—within which there is a Universal Turing Machine, a two-dimensional computer that in principle can compute any computable function. It was far from easy, but they showed how they could "build" a working computer out of simpler Life-forms. Glider streams can provide the input-output "tape," for instance, and the tape-reader can be some huge assembly of eaters, gliders, and other bits and pieces. What this means is mind-boggling: Any program that can run on any computer could, in principle, run in the Life world on one of these Universal Turing Machines. A version of Lotus 1-2-3 could exist in the Life world; so could Tetris or any other video game. So the information-handling ability of gigantic Life-forms is equivalent to the information-handling ability of our real three-dimensional computers. Any competence you can "put on a chip" and embed in a 3-D contraption can be perfectly mimicked by a similarly embedded Life constellation in a still larger Life-form in two dimensions. We know it exists in principle. All you have to do is find it—that is to say, all you have to do is design it.

can exist, just design it

Can We Get the *Deus ex Machina?*

evol, nat sel?

Now it is time to ask whether we might eliminate the miracle-working hacker Gods from our picture, replacing their ingenious design efforts with evolution *within the Life world itself.* Is there any Life world, of any size, in which the sorts of human R&D just described are carried on by natural selection? More precisely, are there configurations of the Life world such that, if you started the world in one of them, it would eventually *do all the work* of the hacker Gods, gradually discovering and propagating better and better avoiders? This move, to an evolutionary perspective, carries with it a family of ideas that can *seem* paradoxical or self-contradictory from our everyday perspective, and it takes some strenuous exercise of thought to get comfortable with the transitions between the two perspectives. One of Darwin's earliest critics saw what was coming and could scarcely contain his outrage:

> In the theory with which we have to deal, Absolute Ignorance is the artificer; so that we may enunciate as the fundamental principle of the whole system, that, IN ORDER TO MAKE A PERFECT AND BEAUTIFUL MACHINE, IT IS NOT REQUISITE TO KNOW HOW TO MAKE IT. This proposition will be found, on careful examination, to express, in condensed form, the essential purport of the Theory, and to express in a few words all Mr. Darwin's meaning; who, by a strange inversion of reasoning, seems to think Absolute Ignorance fully qualified to take the place of Absolute Wisdom in all the achievements of creative skill. (MacKenzie 1868, p. 217)

do not need to know how to make it

ignorance placed wisdom

MacKenzie identifies what he calls a "strange inversion of reasoning," and he is right on all counts. The Darwinian revolution is indeed an inversion of everyday reasoning in several regards, and it is, for that reason, strange: a *foreign* language, full of traps for the unwary, even after considerable practice, all the more so because there are so many terms that are what linguists call *false friends*—terms that seem to be cognates or synonyms of terms from your mother tongue but differ in treacherous ways. One man's *Gift* is another man's poison; one man's *chair* is another man's flesh. (Hint: Look in German–English and French–English dictionaries.) In the case of the Darwinian perspective, the problem of false friends is exacerbated because the terms that invite confusion are, in fact, closely related and relevant to each other—but just not quite the same. When we invert the top-down perspective of tradition and look at creation from the bottom up, we see intelligence arising from "intelligence," sight being created by a "blind watchmaker," choice emerging from "choice," deliberate voting from mindless "voting," and so on. There will be lots of scare-quotes in the explanations to come. We will see—talk about paradox!—how a whole can be more *free* than its parts.

So the straightforward technical question of whether an evolutionary process could replace the effort of the hacker Gods in the Life world has some far-reaching implications. Moreover, the answer has some curious twists in it. In such a Life world, there would have to be self-reproducing entities, and we do know that they can exist, since Conway and his students embedded their Universal Turing Machine in just such a contraption. They devised the Game of Life, in fact, in order to explore John von Neumann's pioneering thought-experiments about self-reproducing automata, and they succeeded in designing a self-reproducing structure that would populate the empty plane with ever more copies of itself, rather like bacteria in a petri dish, each one containing a Universal Turing Machine. What does this machine look like? Poundstone calculates that the whole construction would be on the order of 10^{13} pixels.

> Displaying a 10^{13}-pixel pattern would require a video screen about 3 million pixels across at least. Assume the pixels are 1 millimeter square (which is very high resolution[3] by the standards of home computers). Then the screen would have to be 3 kilometers (about two miles) across. It would have an area about six times that of Monaco.

3. When Poundstone was writing (1985) this was very high, but today it would be low. The pixels on my laptop are almost four times smaller, so the whole screen at that resolution would be somewhat less than 1 kilometer across. Still a big screen.

Perspective would shrink the pixels of a self-reproducing pattern to invisibility. If you got far enough away from the screen so that the entire pattern was comfortably in view, the pixels (and even the gliders, eaters and guns) would be too tiny to make out. A self-reproducing pattern would be a hazy glow, like a galaxy. (Poundstone 1985, pp. 227–28)

In other words, by the time you have built up enough pieces into something that can reproduce itself (in a two-dimensional world) it is roughly as much larger than its smallest bits as an organism is larger than its atoms. That shouldn't surprise us. You probably can't do it with anything much less complicated, though this has not been strictly proven. But self-reproduction is not enough by itself. We also need mutation, and adding this is going to be surprisingly expensive. In his book *Le Ton Beau de Marot* (1997), Douglas Hofstadter draws attention to the role of what he calls *spontaneous intrusions* into any creative process, whether it is achieved by the exertions of a human artist or inventor or scientist, or by natural selection. Every increment of design in the universe begins with a moment of serendipity, the undesigned intersection of two trajectories that yield something that turns out, retrospectively, to be more than a mere collision. We have seen how collision-detection is a fundamental capacity that can be made available to Life-forms, and indeed how collision is a major problem facing all Life hackers, but *how much* collision can we afford in our Life worlds? This turns out to be a serious problem when we set out to add mutation to the self-replication powers of Life configurations.

Computer simulations of evolution abound, and show us the power of natural selection to create strikingly effective novelties in remarkably short periods of time in one virtual world or another, but they are always, perforce, orders of magnitude simpler than the real world, because they are always much more *quiet*. What happens in a virtual world is only what the designer specifies to happen. Consider a typical difference between virtual worlds and real worlds: If you set out to make a real hotel, you have to put a lot of time, energy, and materials into arranging matters so that the people in adjacent rooms can't overhear each other; if you set out to make a virtual hotel, you get that insulation for free. In a virtual hotel, if you want the people in adjacent rooms to be able to overhear each other, you have to add that capacity. You have to add *non*-insulation. You also have to add shadows, aromas, vibration, dirt, footprints, and wear-and-tear. All these non-functional features come for free in the real, concrete world—and they play a crucial role in evolution. The open-endedness of evolution by natural selection depends on the extraordinary richness of the real world, which constantly provides new *undesigned* elements that can be serendipitously harnessed, once in a blue moon, into new design elements. To take the simplest case, can there be enough interference in the world to produce an appropriate number of mutations without, in the process, simply breaking the whole reproductive system? The reproductive system of Conway's Universal Turing Machine was noise-free, making perfect copies every time. There was no provision for mutation at all, no matter how many copies of itself it produced. Could a still larger, more ambitious self-reproducing automaton be designed that could allow for the occasional unblocked glider to arrive, like a cosmic ray, and produce a mutation in the genetic code being copied? Can a two-dimensional Life world be *noisy* enough to support open-ended evolution, while still *quiet* enough to permit the designer parts to do their good work unassailed? Nobody knows.

It is an interesting fact that by the time you specify Life worlds that are complex enough to be candidates for such capacities, they are much too complex to run in simulation. Noise and debris can always be added to a model, but it has the effect of squandering the efficiency that makes computers such great tools in the first place. So there is a sort of *homeostasis* or self-limiting equilibrium here. The very simplicity, the *over*simplicity, of our models can prevent them from modeling the things we are most interested in, such as creativity, either by a human artist or by natural selection itself, since in both cases that creativity feeds on the very complexity of the real world. There is nothing mysterious or even puzzling about this, no whiff of strange new complexity-forces or unpredictable-in-principle emergence; it is simply an everyday, practical fact that computer modeling of creativity confronts diminishing returns because in order to make your model more open-ended, you have to make your model more concrete. It has to model more and more of the incidental collisions that impinge on things in the real world. Encroachment is, indeed, what makes life interesting.

So it is unlikely that we can ever prove *by construction* that somewhere in the Vast reaches of the Life plane, there are configurations that mimic the full open-endedness of natural selection. Still, we can construct the parts piecemeal, providing the important existence proofs we need. Yes, there exist such configurations as Universal Turing Machines, and self-protective persisters, and reproducers, and limited evolutionary processes. Formal arguments such as Wagensberg's (and Conway's and Turing's) take us beyond construction to fill in the gaps of impracticality, so we can say with some confidence that our toy deterministic world is one in which all the necessary ingredients exist for the evolution of . . . *avoiders!* This proposition is what we need to break the back of the cognitive illusion that yokes determinism with inevitability. But before turning to this, it will help to return from toyland to reality, to see what we know about the evolution of avoidance on our planet.

From Slow-motion Avoidance to Star Wars

We know that in the early days—the first few billion years—of life on this planet, self-protective designs emerged, thanks to the slow and non-miraculous process of natural selection. It took on the order of 1 billion years of replication for the simplest life-forms to work out the best designs—still susceptible to revision today, of course—for the basic processes of replication. Along the way there was much *avoidance* and *prevention,* but at a pace much too slow to appreciate unless we artificially speed it up in imagination. For instance, the incessantly exploratory process of natural selection occasionally spewed forth counterproductive DNA sequences, parasitic genes or *transposons,* that hitched a free ride on the genomes of early life-forms, contributing nothing to the well-being of those life-forms but just cluttering up their genomes with extra copies (and copies of copies of copies) of themselves. These parasites created a problem; something had to be *done.* And in due course the incessantly exploratory process of natural selection, by a more or less exhaustive search, "found" a solution (or two, or more): designs for structures in the valuable, constructive parts of genomes that *prevented* the excessive flourishing of these parasites, *counteracting* their *actions* with *reactions,* and so forth. The parasitic genes reacted in turn to this new development by a counterthrust of their own, developed over many hundreds or thousands or millions of gen-

generation

erations, and so it went, and so it continues today. Here the speed limit for avoidance is not the speed of light but the speed of *generation*. The simplest "act" of discrimination—just "noticing" a new problem and getting in position to respond to it—takes a generation, and the trial-and-error process of "figuring out" a solution involves the sacrificial explorations of hordes of variant lineages over many generations. Eventually, though, the good designs emerge victorious—or the lineage perishes, which is the much more likely outcome of all these "efforts" at self-preservation of lineage. A few lucky lineages *happened* to "find" good countermoves. (They weren't *doing* anything, they were just part of what was *happening*—the lucky part, as it happens, that happened to be born with useful mutations.) These lucky ones had descendants, whose descendants—the lucky ones, again—had descendants, and so forth, till you get to us. We—lucky us—are made of such useful parts, exquisitely designed to be good at contributing to avoidance, but now on a much swifter timescale.

And the process continues in the present. Matt Ridley describes the well-studied recent case of the so-called P element, a parasitic "jumping gene" that emerged in a laboratory lineage of fruit flies (*Drosophila willistoni*) in the 1950s, and spread to wild populations of their cousins, *Drosophila melanogaster*.

P element

> The P element has since spread like wildfire, so that most fruit flies have the P element, though not those collected from the wild before 1950 and kept in isolation since. The P element is a piece of selfish DNA that shows its presence by disrupting the genes into which it jumps. Gradually, the rest of the genes in the fruit fly's genome have fought back, inventing ways of suppressing the P element's jumping habit. (Matt Ridley 1999, p. 129)

no noticer, decider

How long did it take these genes to "recognize" the problem and "fight back"? Many generations, but notice that there was no central noticer, no decider. What happened is just what always happens in natural selection. The impact of the P elements was not uniform on all lineages of fruit flies; there was variation in the genomes of fruit flies, some of which were better able to cope with this new challenge. Those that coped prospered, and then those of their offspring that coped even better prospered even more, so that in due course "solutions" to the problems posed by those P elements emerged and were "discovered" and "endorsed" by Mother Nature, otherwise known as natural selection. It can't happen any faster than that in nature; the exploration cannot precede the posing of the problem (that would be evolutionary precognition), and thereafter each step takes a whole generation at the least. Fortunately, the exploration can take advantage of "parallel processing" by exploring in all actual (though not all possible) lineages of fruit flies at once, so that the problem-solving can happen quite swiftly, in less than a half-century in the case of fruit flies.

One of the standard (and much-needed) correctives issued to those who study evolution is the old line about how natural selection has no foresight at all. It is true, of course. Evolution is the blind watchmaker, and we must never forget it. But we shouldn't ignore the fact that Mother Nature is well supplied with the wisdom of hindsight. Her motto might well be "If I'm so myopic, how come I'm so rich?" And while Mother Nature is herself lacking in foresight, she has managed to create beings—us human beings, preeminently—who do have foresight, and are even beginning to put this foresight to use in guiding and abetting the very processes of natural selection on this planet. I occasionally encounter even quite sophisticated evolutionary theorists who find this paradoxical. How could a process with no foresight invent a process with foresight? One of the main goals of my book *Darwin's Dangerous Idea* was to show that this is not paradoxical at all. The process of natural selection, slowly and without foresight, invents processes or phenomena that speed up the evolutionary process itself—cranes, not skyhooks, in my fanciful terminology—until the souped-up evolutionary process finally reaches the point where explorations within the lifetime of individual organisms can affect the underlying slow process of genetic evolution, and even, in some circumstances, usurp it.

Today we human beings can see and hear things at a distance, and don't have to wait for them to sidle up to us. Thanks to our long-distance perceptual organs and our prosthetic extensions of them, we can pose and solve problems at a tempo approaching the maximum speed limit of the physical universe: the speed of light. Anything faster than that would be precognition, which we can't do, but we actually do butt right up against the speed of light in our problem-recognition and problem-solving capacities. Thanks to our technology, for example, we can detect the liftoff of a nuclear missile within milliseconds of its occurrence thousands of miles away, and then use that precious lead time to design a countermeasure that has some non-zero chance of working. It's a breathtaking feat of avoidance, of dodging an incoming brick. (Can we really? Haven't I myself often claimed that Ronald Reagan's Strategic Defense Initiative and its descendants—often called Star Wars—are a technologist's fantasy, systematically incapable of successful implementation? But if Star Wars is currently impossible, as I do indeed believe, that is only because it is at the cutting edge of the avoidance arms race *today,* and the readily imaginable countermeasures seem to have the upper hand; they would almost certainly succeed in *preventing* the *prevention* that is the goal of Star Wars, though surely many of the missiles would be successfully intercepted, which is all that I am claiming. I am not a fan of Star Wars, but I am nevertheless delighted to find that this criminally expensive and irresponsible system can be put to some modest use after all, if only as a philosopher's example!)

We are virtuoso avoiders, preventers, interferers, forestallers today. We have managed to get ourselves into the happy situation of having enough free time to sit around systematically looking into the future and asking ourselves what to do next. We are squeezing every drop of information out of the world that we can, and then we massage it all into breathtaking new vistas onto what will come. And what do we see? We see that there are *some* inevitabilities, but actually the list gets shorter every week. It used to be that there was nothing we could do about tidal waves, or flu epidemics, or hurricanes (we can't yet deflect them, but we have plenty of advance warning so that we can hunker down and minimize the damage). It used to be that a person who fell out of a boat in the dead of night in the middle of the ocean was a goner for sure. Now we can fly in helicopters guided by homing devices and pluck people out of the deep for all the world like the phony miracles of the old *deus ex machina* of Greek drama. This is all a very recent biological development. For billions of years there was nothing like it on the planet. Processes were either entirely blind or at best myopic, clueless and reactive, never foresightful or proactive.

As we have seen, it is easy for us, inveterate and imaginative agents that we are, to discern the pattern of avoidance and prevention at many different timescales, from the supersonic to superglacial. We can effortlessly extend it to atoms and even subatomic particles if we like, thinking of them as if they too were tiny agents, worrying about their own futures, hoping to contribute to some great campaign, persisting as best they can in the world of hard knocks. We can imagine, if we like, that atoms cringe just before the anticipated collisions occur. That would be silly, of course. Atoms have no foresight, no interests, no hopes; they are just tiny places where there is *happening,* not *doing.* But that doesn't stop us from simplifying our vision of them by treating them as if they were agents—very simple, single-minded agents. That carbon atom clings tenaciously to those two oxygen atoms, *preventing* them from wandering off, forming a persistent molecule of carbon dioxide—a modest task for a carbon atom. Other carbon atoms play more exciting roles holding together gigantic mega-atom proteins, so that the proteins can *do their thing,* whichever thing it is.

I suspect that we find it natural to keep track of the complexities of atoms and the stranger denizens of the world of subatomic physics by treating them rather like tiny agents because our brains are designed to treat everything we encounter as an agent if possible—just in case it really is one. In the early days of human culture, the childhood of civilization, you might say, we found it useful to overuse this *animism,* treating all of nature as made of gods and fairies, malevolent and benevolent sprites, imps and goblins in charge of all the natural processes we observed. It was intentional systems all the way down, you might say. This tactic has been moderated and sophisticated—ever since Democritus, in fact—so that now we are quite comfortable thinking of atoms as just little mindless bouncing grains. They don't quite *act* but they still *do things*—repelling and attracting, wobbling in one place or dashing off.

I am not suggesting that there is a clear-cut division, in the end, between things that merely happen and things that do things, valuable as that opposition is. As usual, we get a grading off of instances from florid to pastel to invisible, a diminution of the appropriateness of the family of concepts anchored to our predicament as agents trying to preserve ourselves. After all, an avalanche can destroy a village and kill people just as surely as a marauding army can, and even simple helium atoms can push against the inside of a balloon, keeping it stretched taut. Yes, and enzymes can be busy little agents indeed. I suspect, in fact, that it is our *inability* to make ready sense of subatomic events in such familiar agency terms that makes the world of subatomic physics such an alien and hard-to-conceive arena of events. The familiar concepts of cause and effect, as we shall see in the next chapter, are much better anchored to our macroscopic world of agency than to the underlying world of microphysics.

The Birth of Evitability

It is time to take stock and consider some objections that I have postponed. The main point of this chapter is to show that we need to take the etymology of "inevitable" seriously. It means *unavoidable*. Curiously, its negation is not used,[4] but we can easily enough coin the term, and note that some things are *evitable* by some agents, and some things, in contrast, are not evitable by those agents. We have seen that in a deterministic world such as the Life world we can design things that are better at avoiding harms in that world than other things are, and these things owe their very persistence to this prowess. Of all the things we see on a particular Life plane, which will still be there a billion time steps from now? The harm-avoiders have the best chance. We can put the main point of the chapter as the conclusion of an explicit argument:

In some deterministic worlds there are avoiders avoiding harms.
Therefore in some deterministic worlds some things are avoided.
Whatever is avoided is avoidable or evitable.
Therefore in some deterministic worlds not everything is inevitable.
Therefore determinism does not imply inevitability.

This argument seems a bit fishy, doesn't it? That's because it exposes hidden assumptions about avoiding and inevitability that have gone largely unnoticed. Pointing to particular instances of avoidance as proof of "evitability" seems odd because it runs contrary to a typical way of thinking about inevitability:

If determinism is true, then whatever happens is the *inevitable* outcome of the complete set of causes that obtain at each moment.

This may be a familiar way of speaking, but what does it mean? Compare it with the trivially true claim:

---

4. The *Oxford English Dictionary* lists "evitable" as a word recorded in 1502, marking it as obsolete, except in the negative.

If determinism is true, then whatever happens is the *determined* outcome of the complete set of causes that obtain at each moment.

If "inevitable" is not just a synonym for "determined," what does it additionally convey? Inevitable outcome? Inevitable by whom? Inevitable by the universe as a whole? That makes no sense, since the universe isn't an agent with an interest in avoiding anything. Inevitable by anybody? But that is false; we've just seen how to distinguish the skillful avoiders from their less talented kin in some deterministic worlds. When we say that some particular outcome is inevitable, we might mean that it is inevitable by all the agents living at that time and in that place, but whether or not this is true is independent of determinism. It depends on the circumstances. This all needs some further unpacking, and who better to help me than Conrad, your ombudsman.[5]

> **CONRAD:** The configurations in the Life world that happen to—that *seem* to—avoid this and that are not *really* avoiding anything, of course. After all, each of them "lives" in a deterministic world, and if you rerun the tape a million times, each of them will "do" exactly the same thing— exactly the same thing will happen—no matter how much "evolution" has gone on in that world. In the Life world evolution scenario, each particular avoider, situated on the plane exactly where it is, *comes to the particular fate it was always destined to come to*—it either avoids harm until after it replicates or it doesn't. If it confronts a thousand "avoidance" opportunities before it's killed off, that's exactly the life it was always going to have. You speak above of the avoiders "having the best chance" of surviving, but, of course, chance doesn't enter into it! Those that survive survive and those that don't don't, and that's all determined from the outset.

5. Conrad is the cousin of Otto, the fictional articulator of various objections and challenges to my theory of consciousness in *Consciousness Explained*. Otto has been variously described in reviews as my "stooge" and my "conscience," but for better or worse he expressed as vividly and sympathetically as I could muster the most common misgivings I encountered to my views on that topic. Everything Conrad says in this book is a distillation of, and—so far as I can manage it—an improvement on, the most common and pressing objections and misgivings I have encountered to the claims in this book. He often speaks for the critics I thank in the Preface, and if I have calculated correctly, you will find that he often speaks for you.

chance

As we'll see in the next chapter, there is a perfectly good concept of *chance* that is compatible with determinism, and it's the concept we invoke to explain evolution, among other things. (Evolution doesn't depend on indeterminism.) But, meanwhile, you are right that each trajectory in the Life world is perfectly determined, but why do you insist that determined avoidance isn't real avoidance? The long-term process of which each such simple avoider (or pseudo-avoider, if you insist) forms a mindless part, just happening along and playing out its "destiny," has a remarkable power: It gradually produces better and better (pseudo-) avoiders, more and more adroit copers with Life's problems—though, of course, the problems become more severe, too; it's a rat race. The fact that the whole process is determined doesn't detract from the fact that as time passes it generates more and more of something that looks for all the world like avoidance.

looks like avoidance

> **CONRAD:** It may look like avoidance, but it's not *real* avoidance. *Real* avoidance involves changing something that was going to happen into something that doesn't happen.

I guess it all depends on what you mean by "going to happen." Are you perhaps being misled by the simplicity of the imaginary examples in the Life world? There is a contrast between simple, "hard-wired" avoidance responses and fancier varieties, but you can't use it to contrast real-world avoidance with Life-world avoidance. A nice example is the blink reflex, which is tuned on a hair trigger in us, so that most of the time when we blink in response to a swiftly looming something, it is a false alarm. No incoming debris was destined for our eyes after all; there was nothing for our eyelids to form a temporary wall against. In the trade-off between the costs of wasting energy and interrupting one's vision briefly, and the costs of passing up an opportunity to blink that would have saved an eye, Mother Nature has "erred on the side of caution," probably because the costs (in time and energy) of getting more information before committing to action rise much too steeply. Blinks are, in general, *involuntary,* but other reactions can be suppressed. The human brain devotes an elaborate subsystem to analyzing motion-in-depth,

> with the lion's share of representational space devoted to the cone of directions which intersect the head. Again, the rationale for this representational scheme is intuitively obvious—we are most

"interested" in objects that are quickly approaching our heads. Intuitively, that is, it is the baseball that's going to hit you smack in the face that's of interest, not the ball about to clear your left shoulder—and the representational system reflects this fact. (Akins 2002, p. 233)

avoidance system

But in what sense was that baseball "going to" hit you smack in the face? You dodged it; you were *caused* to dodge it by the elaborate system evolution has built into you to respond to photons bouncing off incoming missiles on certain trajectories. It was "never really going to" hit you precisely because it caused your avoidance system to go into action. But that avoidance system is more sophisticated than a simple blink reflex, and it can respond to further information, when it is available, and countermand its initial decision. Noticing that you can win the game for your team by being hit by the incoming pitch, you can decide to take the hit. You avoid doing the avoiding that was well within your power—thanks to (*caused by*) the advance notice you had of the wider context. And you can also avoid avoiding avoiding, when circumstances warrant it. This open-ended human ability is a far cry from the simple harm-ducking configurations we've imagined in the Life world, but if you're tempted to think that only simple, "hard-wired reflexes" (mere pseudo-avoidance, one might call it) can evolve in the Life world, you're mistaken. All the layers of sensitivity and reflection we human beings exhibit are accessible in principle to Life configurations. After all, there are Universal Turing Machines in the Life world.

> **CONRAD:** I take your point, but I still think that what happens in the Life world, of whatever complexity or sophistication, doesn't count as genuine avoidance, which involves actually *changing the outcome*. Determined avoiding isn't real avoiding because it doesn't actually change the outcome.

determined avoiding?

From what to what? The very idea of changing an outcome, common though it is, is incoherent—unless it means changing the *anticipated* outcome, which we've just seen is exactly what happens in determined avoiding. The *real* outcome, the *actual* outcome, is whatever happens, and nothing can change *that* in a determined world—or in an undetermined world!

nothing can chg what actually happens

CONRAD: But still, those entities in the Life world that have these various powers of so-called avoidance _inevitably_ have just the powers they have and are _inevitably_ placed in the world just where they are, at all times, thanks to the determinism of that world, and the initial position in which it starts.

No, this is precisely the use of "inevitably" that I am calling into question. If all you mean is that the powers each of them has to avoid things is _determined by the past,_ then you are right, but you must break this bad habit of yoking determinism with inevitability. _That_ is the reflection that needs to be disabled at the outset, for if it doesn't apply to your dodging—or not dodging—the baseball, then it also doesn't apply to the many apparent feats of avoidance exhibited by simpler dodgers in the deterministic Life world. If we want to make sense of the biological world, we need a concept of avoidance that applies liberally to events in the history of life on Earth, whether or not that history is determined. This, I submit, is the _proper_ concept of avoidance, as real as avoidance could ever be.

It is worth noting, finally, that just as evitability is compatible with determinism, inevitability is compatible with indeterminism. Something is inevitable _for you_ if there is nothing _you_ can do about it. If an undetermined bolt of lightning strikes you dead, then we can truly say, in retrospect, that there was nothing you could have done about it. You had no advance warning. In fact, if you are faced with the prospect of running across an open field in which lightning bolts are going to be a problem, you are much better off if their timing and location are determined by something, since then they _may_ be predictable by you, and hence avoidable. Determinism is the friend, not the foe, of those who dislike inevitability.

This should serve to break the traditional, or perhaps habitual, link between determinism and despair. There are other familiar habits of thought that should also be broken, or at least set aside for skeptical scrutiny. To speak about prevention or avoidance in the pre-biological or a-biological universe is to project a concept beyond its home base in our manifest image as agents, not always in illusory ways, but at least with the prospect of opening up unwanted implications. How much prevention is there in our world? We speak of gravity preventing the underpowered rocket from entering orbit, because this is a topic that interests us. We are less likely to speak of gravity preventing the beer in a glass from floating around the room, but not because it is any less reliable a regularity. As you read this, your beating heart is postponing your death, and your attention to the page is preventing you from seeing all manner of other things in your immediate environment. You may well be avoiding a sprained ankle by not walking at this time, but also hastening the decay of the chair you are sitting in. We can easily conjure up scenarios in which these regularities get dramatized as cases of prevention, enabling, thwarting, deflecting, undoing, counteracting, and the like, and this is often a useful perspective to adopt toward these regularities, but the habit of thought or policy should be recognized for the anthropocentric (or at least agentocentric) projection that it is.

> **CONRAD:** OK. I see that I can't just help myself to the term "inevitable" in the standard way, but I still have a strong suspicion that you're pulling a fast one on me. I think there must be *some* sense of "inevitable" in which what happens in a determined world is inevitable. And I don't see anything that looks like what *I* call free will happening in the Life world.

Fair enough. We'll keep looking in later chapters for that elusive sense of "inevitable," but you agree that in the meantime I've shifted the burden of proof: There shall be no inferring inevitability *in any sense* from determinism without mounting a supporting argument. And I agree that we are still a long way from free will. There is nothing that looks remotely like freedom at the level of the physics of the Life world. Gliders and eaters aren't the slightest bit free, and what they do is what they have to do, every time. It seems to stand to reason that nothing composed of such unfree parts could have any more freedom, that *the whole cannot be freer than its parts,* but this hunch, which is the very backbone of resistance to determinism, will turn out, on closer examination, to be an illusion. In the next chapter, we will look at this agent's-eye vision of cause and effect, possibility and opportunity, more closely, to see in more detail why the important issue of inevitability has nothing whatever to do with the question of determinism.

then — nothing to do w/ determinism

ev- in det. world
inev-@design level

Chapter 2

A toy model of determinism demonstrates that in the Vast space of possible con-figurations of "matter" there are some that persist better than others, because they have been designed to avoid harm. The process by which these entities emerge uses information gleaned from the environment to anticipate general and sometimes particular features of likely futures, permitting informed guidance. This proves that evitability can be achieved in a deterministic world, and hence that the common association between determinism and inevitability is a mis-take. The concept of inevitability, like its source concept of avoidance, properly belongs at the design level, not the physical level.

Chapter 3

The concepts of causation and possibility lie at the heart of anxiety about free will, and an analysis shows that our everyday concepts do not have the impli-cations they are often assumed to have: Determinism is no threat to our most important thinking about possibilities and causes in our lives.

Notes on Sources and Further Reading

There are more extended arguments for the conclusions drawn in this chapter in my "Real Patterns" (1991B), *Darwin's Dangerous Idea* (1995), *Kinds of Minds* (1996A), and, most recently, "Collision Detec-tion, Muselot, and Scribble: Some Reflections on Creativity" (2001A).

A "simple" Life world Turing machine, expandable (in imag-ination, not practically) into a Universal Turing Machine, has been executed by Paul Rendell, and can be seen and explored at his Web site: http://www.rendell.uk.co/gol/tm.htm. His list of parts—all crafted from gliders, eaters, and their kin—is inspiring: 1Gap3, 1Gap4, 1Gap8, Column Address, Comparator, Control Conversion, Fanout, Finite State Machine, In Gate, Memory Cell, Metamorphosis II, MWSS Gun, Next State Delay, NOT XOR Gate, Outgate, Output Collator, P120 Gun, P240 Gun, P30LWSS Gun, P30MWSS Gun, Pop Control, Push Control, Row Address, Set Reset Latch (a), Set Reset Latch (b), Signal Detector, Stack, Stack Cell, Takeout, Turing Tape.

Chapter 3

THINKING ABOUT DETERMINISM

Determinism seems to rob us of our opportunities, seems to seal our fates in the total web of causal chains extending back into the past. We generally ignore this dire prospect. We all spend quite a lot of time thinking about how things *may* go today or next year, or *might* have gone if only such and such. We seem, in other words, to assume that our world is not deterministic.

Possible Worlds

We readily distinguish in our deliberations between ways things could have gone and ways things couldn't have gone, between how things won't go no matter what happens and the way things may well go, if we so choose. As philosophers say, we often imagine *possible worlds:*

> In World A, Oswald's shots missed Kennedy and hit LBJ instead, changing subsequent history in millions of ways.

And we use these imaginings to guide our choices of action, although only a philosopher would be apt to put it that way:

> I just imagined a world just like the actual world except that I didn't eat that eclair and hence didn't experience the regret I'm now feeling.
>
> In World A, I propose to Rosemary. In World B, I send her this farewell note I'm writing and join a monastic order.

> Familiar as this exercise of the imagination is, it often plays tricks on us when we try to think rigorously about determinism and

poss ible - compatible w/ determ.

causation. In this chapter, I will argue that determinism is entirely compatible with the assumptions that govern our thinking about what is possible. The apparent incompatibility is a cognitive illusion, plain and simple. There is no such conflict. Both in our everyday thinking about what to do next, and in our most careful scientific thinking about the causes of phenomena, we employ concepts of *necessity, possibility,* and *causation* that are strictly neutral with regard to the question of whether determinism or indeterminism is the truth. If I am right, then more than a few eminent philosophers are wrong, so expect some heavy artillery—but rumbling in the distance, since I am not going to do direct battle with them here. Christopher Taylor has greatly clarified my thinking on this topic and shown me how to launch a deeper and more radical campaign in support of my earlier claims to this effect, and our coauthored paper (Taylor and Dennett 2001) provides more technical detail than is needed here. Here I will attempt a gentler version of our argument, highlighting the main points so that nonphilosophers can at least see what the points of contention are, and how we propose to settle them, while leaving out almost all the logical formulae. Philosophers should consult the full-dress version, of course, to see if we have actually tied off the loose ends and closed the loopholes that are passed by without mention in this telling. And since what follows in this telling is due in large measure to Taylor, there will be a temporary shift in authorial pronouns to "we."

Our task, then, is to clarify the *everyday* concepts of possibility, necessity, and causation that arise in our thinking, our planning, our worrying, our imagining, as we cope with the world and its challenges. We can simplify our task by restricting our thinking about possible worlds to thinking about Quine's Democritean universes. Quine was famously skeptical about all attempts to speak seriously about possibility and necessity—the topic of *modal logic*—and his Democritean universes were concocted in order to provide a maximally tame, orderly base of operations from which the issues could be explored. As you will recall from Chapter 2, each of the Vastly many Democritean universes consists of a swarm of point-atoms whose trajectories through space and time are given by their four-dimensional coordinates $\{x, y, z, t\}$. A complete *state description* of the world at time t is simply the exhaustive catalogue of the occupied addresses $\{x, y, z,\}$ at t. We call the set of all *logically* possible worlds the Library of Democritus, and let's call the subset that contains just the *physically possible* worlds Φ—phi. Of course, we don't yet know all the laws of physics, and don't know for sure whether they are deterministic or indeterministic, but we can pretend that we know them. (Now that we have Conway's Life world under our belts, we can always check our intuitions by recasting the issue into Conway's Life world, where we *do* know the physics perfectly and know that it is deterministic.)

Given a possible world, we have many ways to make assertions about it. As we saw in the case of the simple world of Life, it will typically be natural to leap up above the atomic level and describe the world in terms of larger chunks of stuff. Just as we could trace the career of some particular glider from its birth to its death on the Life plane, so we can track the trajectories through time and space of such "connected hypersolids" (four-dimensional objects) as stars, planets, living creatures, and everyday paraphernalia—the familiar objects found in human lives. Plato speaks in a famous image of carving nature at its joints, and the joints *we* start with—literally, where one *thing* leaves off and the next *thing* starts—are the patterns that are salient and stable enough *for us* to identify (and track, and reidentify) as macroscopic things. As we saw in the Life world, the underlying "physics" (the state transition rule) dictates which configurations are robust enough over time to constitute macroscopic (not microscopic) regularities, and we use these to anchor our imaginations when we think about causes and possibilities. We can describe such middle-size patterns of atoms using the familiar system of *informal predicates* that apply to these entities, such as (in order of increasing contentiousness) "has a length of 1 meter," "is red," "is human," "believes that snow is white." These informal predicates unleash a horde of problems concerning vagueness, subjectivity, and intentionality, and it is these problems—the problems that arise when you leap up from the basic level of atoms and space to higher ontological categories—that fueled Quine's skepticism about the likelihood of making sense of talk about possibility and necessity. We think that by highlighting the move, and concentrating all the slippage into the move from the atomic physical level to the everyday level, we can keep these problems isolated so that they do not imperil our basic approach. Proceeding gingerly, then, and assuming that we can get some tentative grip on informal predicates, we may then in good conscience form sentences like

(1) There is something that is human.

and determine whether they apply in various different possible worlds. There are no human beings in any Life world, since human beings are three-dimensional beings, but there may be two-dimensional entities that are wonderfully reminiscent of human beings in some of them. Closer to home, would a possible world in which the language-using, technology-exploiting, culture-creating bipeds had feathers instead of hair on their heads and had descended from ancestors of ostriches be a world in which there was something that was human? Or would we call such a creature a non-human *person?* Is "human" a biological category or, as the word "humanities" suggests, a sociocultural or political category? Opinions may differ on how to interpret the informal predicate "human." Often enough one will encounter borderline worlds where incontestable verdicts prove elusive.

Worthy of special note are *identification predicates* of the form "is Socrates." "Is Socrates," we shall suppose, applies to any entity in any possible world that shares so many features with the well-known denizen of the actual world that we are willing to consider it the same person. In the actual world, of course, "is Socrates" applies to exactly one entity; in other worlds, there may reside no such being, or one, or conceivably two or more to whom the predicate applies equally well. Like other informal predicates, identification predicates suffer from vagueness and subjectivity, but these vexatious issues can be isolated and dealt with as they arise in particular cases.[1]

Now we are ready to define the fundamental concepts we need—*necessity, possibility,* and *causation*—in terms of possible worlds. Such a sentence as

(2) Necessarily, Socrates is mortal.

we may translate as

(3) In every (physically?) possible world *f,* the sentence "If anything is Socrates, it is mortal" is true.

---

1. Maven alert: Yes, we are sidestepping the battles over rigid designation, at our peril. Catch us if you can. (Rigid designation is a concept due to Kripke [1972], and opinion is divided over whether it succeeds in resurrecting essentialism. We think not, but would rather not spend the rest of the year defending our view.)

In other words, when we cast our minds around canvassing all the possibilities we can contemplate, we find there is not a single possible world that has an immortal Socrates in it. *That's what it means* to say that Socrates is necessarily mortal. Here "is Socrates" and "is mortal" are informal predicates of the sort just introduced. Deciding whether the sentence is true does present many challenges, of course, stemming in large part from the unavoidable blurriness of the predicates: Is a Socrates-candidate that is mortal but can fly like Superman less worthy of the predicate "is Socrates" than a Socrates-candidate that is earthbound but miraculously unaffected by his cup of hemlock? Who's to say? Moreover, we haven't yet decided whether the set of possible worlds over which we should allow *f* to range should be the whole Library of Democritus (all worlds) or Φ (the physically possible worlds) or even some still more restricted set X. Logic alone can't resolve this issue, but logical language does help us to pinpoint such questions and discover more precisely the sorts of vagueness we face.

Now we can define *possibility*. What is *possible* is whatever isn't *necessarily not* the case, so

(4) Possibly, Socrates might have had red hair.

means

(5) There is a (at least one) possible world *f* in which the sentence "There is something that is Socrates and he has red hair" is true.

Once again, we have to decide whether this is physical or logical possibility we are talking about. It is *physically* possible if there is a world *in set* Φ with a red-haired Socrates. Otherwise, this is ruled out, physically, no matter how common red-haired Socrates is in logically possible but physically impossible worlds.

Now we are in a position to clarify the definition of determinism given at the beginning of Chapter 2: *There is at any instant exactly one physically possible future*. To say that determinism is true is to say that our actual world is in a subset of worlds that have the following interesting property: There are no two worlds that start out exactly the same (if they start the same, they stay the same forever—they are not *different* worlds at all), and if any two worlds share *any* state description exactly, they share all subsequent state descriptions. The Life world illustrates this crisply. It is deterministic in only one direction; you cannot in general extrapolate the *previous* instant the way you can always extrapolate the *next* instant. For instance, a Life plane containing a single square-of-four still life at time *t* (see Figure 3.1) has an ambiguous past. The next state (and the next, and so forth) is exactly the same—unless something encroaches—but the previous state could have been any of these five (or indefinitely many others with more distant evaporating ON pixels).

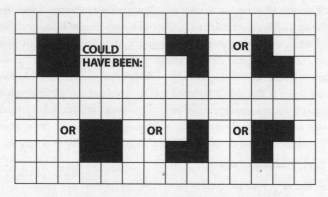

Figure 3.1   Still Life and What Could Have Been

So if determinism—thus defined—is true, we can conclude that even if many different pasts *might* have led to exactly our present state, our future is "fixed" by our present state. From this perspective, determinism *seems* to be just the opposite of our standard vision, in which the past is "fixed" and the future is "open." We could define a stronger (and non-standard) form of determinism that excludes such ambiguous pasts, ruling out what I have called *inert historical facts*—facts about the past that, so far as the laws of physics are concerned, could have been one way or another without leaving any subsequent effect. The ability of cosmologists to "run the movie backward" and thereby calculate facts about the early moments after the Big Bang shows that with regard to some properties, we can read the past off the present with stunning precision and reach, but this goes no way at all to show that there are no inert historical facts. The fact that some of the gold in my teeth once belonged to Julius Caesar—or its negation, the fact that none of it ever did—is a plausible example of an inert historical fact.

It is certainly *practically* inert. Since we don't happen to keep track of
the chain of ownership of bits of gold the way we do for, say, paint-
ings by Rembrandt, it is well-nigh unimaginable that any investigation
of the current state of the world's distribution of atoms would enable
one to figure out which of these two sentences is true, but one of them
surely is.

And when we look into the future, it is next to impossible to
tell when a *heretofore* inert historical fact will emerge to "make a difference" in what happens next. Suppose determinism is true and we
know the laws of physics perfectly, the way Laplace's demon knows
them. Still, unless we have *perfect and complete* knowledge of one state
description of the universe, we will not be able to tell which of Vastly
many microscopically different *possible* worlds in the set Φ is the *actual*
world. It is because our knowledge is inevitably incomplete that thinking in terms of possible worlds is such a good fallback.

One of the most useful applications of possible-world talk is in
interpreting *counterfactual* sentences, such as

(6) If Greenspan had sobbed in Congress, the market would have
crashed.

and

(7) If you had tripped Arthur, he would have fallen.

Following David Lewis (1973), we can see that (*roughly*) sentence (7)
is true if and only if in every world *approximately similar* to our own
where the *antecedent* holds, so does the *consequent*. In other words,

(8) Take the set of worlds X similar to our actual world: In each
world in that set where there is an instance of *you tripping Arthur*,
there is also an instance of *Arthur falling*.

Sometimes when we make counterfactual claims like this, we actually
find ourselves checking them by *imagining* a few variations along these
lines ("Let's see, suppose Arthur was wearing a red shirt, would *that* have
kept him from falling? Suppose the radio had been turned down, suppose the heat was turned off, suppose he was wearing knee pads. . . .
No, he still would have fallen. Suppose the room was filled with inflated
air bags or the whole building was in free fall with zero gravity, now
that would have kept him from falling. . . . But that is too dissimilar to

count"). And in controlled experiments, we don't just imagine, we actually investigate the variations. We vary the conditions systematically, seeing what changes and what doesn't. This is not as straightforward as it first appears, as we shall see.

Whether or not we conduct any actual experiments or thought experiments, what we imply by asserting the counterfactual is that some such set of worlds X similar to our actual world has this regularity in it. And in general, we can express the interpretation of a counterfactual like (6) or (7) as

(9) In the set of worlds X, A $\Rightarrow$ C,

where A is the antecedent and C is the consequent.

But *how similar* to our world should the worlds in set X be? Choosing an optimal value for X in these cases is not always easy, but we can follow loose Guidelines:

In sentences like (6) and (7), X ought to:

- contain worlds in which A holds, not-A holds, C holds, and not-C holds

- contain worlds otherwise very similar to the actual world (insofar as the preceding clause permits).

So when analyzing (7), choose X to contain worlds in which you trip Arthur, worlds where you refrain from tripping him, worlds where he falls, and worlds where he remains upright. (Notice how we use our higher-level ontologies to gather these similar worlds together. We *don't* grade the similarity of worlds by counting how many different voxels there are filled with iron or gold; we use the informal predicates, with all their slop and vagueness, to determine which worlds to include. It turns out, as we shall see, that many of the quandaries that arise for claims about causation and possibility hinge on how we choose set X, the comparison set of nearby possible worlds.)

Causation

Finally, what about *causation?* Some philosophers hope someday to unearth the one "true" account of causation, but given the informal, vague, often self-contradictory nature of the term, we think a more realistic goal is simply to develop a formal analogue (or analogues) that helps us think more clearly about the world. Our preexisting hunches about causation will provide some guidance, but we should mistrust any informal arguments that masquerade as "proofs" validating or debunking particular causal doctrines.[2] When we make an assertion like

(10) Bill's tripping Arthur caused him to fall.

a number of factors appear to be at work supporting the claim. In an approximate order of importance, we list the following:

- Causal necessity. Our assent to sentence (10) depends on our conviction that had Bill not tripped Arthur, he would not have fallen. Using the interpretation of counterfactuals just given, we choose set X, the set of worlds similar to our own, as a set including worlds in which (i) Bill trips Arthur; worlds in which (ii) Bill doesn't trip him; worlds in which (iii) Arthur falls; and worlds in which (iv) he doesn't fall. And we check to make sure that in this set X, in all the worlds where Arthur falls, Bill tripped him.

- Causal sufficiency. It may well be that whenever we affirm (10), we do so partly because we believe that Arthur's fall was an *inevitable outcome* of Bill's tripping: In *any* world where Bill places the obstruction in his path, Arthur goes toppling. (There's that word "inevitable," and it does mean *unavoidable* here: Arthur—for one reason or another—cannot *avoid* falling, and Arthur's friends cannot *prevent* him from falling, and there's nothing else in the offing to *interfere with* his falling, and so forth; gravity will not be defied on this occasion.) This second condition is logically entirely distinct from the first, and yet the two seem to get badly muddled in everyday thinking. Indeed, as we shall see, confusion often originates precisely

---

2. These are fighting words to some philosophers, of course. Fine; we happily shift the burden of proof to them. If they can come up with some unproblematic, counter-example-free theory of the *whole* ordinary concept of causation, we will then compare our more modest, sketchy project to it and see whether we've left out anything important. Meanwhile, we can get on with our analysis using our partial account of what strikes us as the most important aspects of the everyday concept.

here. Below we will discuss at greater length the relations between these two conditions.

- Independence. We expect the two sentences A and C to be logically independent. That is, in possible-world terms, there must exist worlds, however remote from reality, in which A obtains but not C, and vice versa. Hence "Mary's singing and dancing caused her to dance and sing" has a decidedly odd ring. This condition also helps rule out "$1 + 1 = 2$ causes $2 + 2 = 4$."

- Temporal priority. A reliable way to distinguish causes from effects is to note that causes occur earlier. (Maven alert.)

- Miscellaneous further criteria. Although less critical than the preceding points, a number of other conditions may increase our confidence when we make causal judgments. For instance, in textbook examples of causation, A often describes the actions of an agent, and C represents a change in the state of a passive object (as in "Mary causes the house to burn down"). Further, we often expect the two participants to come into physical contact during their transaction.

In order to understand these conditions better, let's try them out on a few test cases, some of which derive from Lewis (2000). First consider the sharpshooter aiming at a distant victim. Suppose that scrutiny of the sharpshooter's past record shows that the probability of a successful hit in this case is 0.1; if you think it makes any difference, we might imagine that irreducibly random quantum events in the intervening air, or in the sharpshooter's brain, help determine the outcome. Let us suppose that in the current case the bullet actually hits and kills the victim. We unhesitatingly agree then that the sharpshooter's actions *caused* the victim's death, *despite their causal insufficiency.* Accordingly, it appears that *at least in cases like these,* people rank necessity above sufficiency when making judgments about causes.

Still, sufficiency does retain some relevance. Suppose that the king and the mayor both have an interest in the fate of some young dissident; as it happens, both issue orders to exile him, so exiled he is. This is a classic case of *overdetermination.* Let A_1 stand for "the king issues an exile order," A_2 stand for "the mayor issues an exile order," and C, "the dissident goes into exile." In this scenario, neither A_1 nor A_2 alone is necessary for C: For instance, had the king failed to issue any order, the dissident would still have been exiled thanks to the mayor, and vice versa. Instead, sufficiency comes to the rescue and permits a choice between the two. In this instance A_2 fails the test: It is easy to imagine a universe where the mayor issues his decree, yet the dissident gets off (just change the king's order into a pardon). The king's order, on the other hand, is truly *effective;* whatever small changes we make to the universe (including changes in the mayor's orders), the dissident's exile follows from the king's command. Accordingly we may dub A_1 the "real cause" (if we feel the need to satisfy that yearning).

Finally, consider the tale of Billy and Susie. Both children are throwing rocks at a glass bottle, and, as it happens, Susie's rock, traveling slightly faster, reaches the bottle first and shatters it. Billy's rock arrives a moment later at exactly the spot where the bottle used to stand, but of course encounters nothing but flying shards. When choosing between A_1 ("Susie throws rock S") and A_2 ("Billy throws rock B"), we vote for A_1 as the cause of C ("The bottle shatters"), despite the fact that neither sentence is necessary (had Susie not thrown her rock, the bottle would still have shattered thanks to Billy, and vice versa) and both are sufficient (Billy's throw suffices to produce a broken bottle, whatever his playmate does, and likewise with Susie's). Why? The general notion of temporal priority (introduced above in connection with distinguishing cause from effect) strikes us as one critical consideration. As with priority disputes in science, art, and sports, we seem to put a premium on being the *first* with an innovation, and since rock S arrived in the vicinity of the bottle earlier than rock B, we give credit to Susie. Further, it is clear that, although the bottle would still have shattered without Susie's throw, the shattering event would have been significantly different, occurring at a later time with a different rock sending fragments off in different directions. (Notice that this problem arises precisely because we've leaped up to the everyday ontology of bottles and breakings, and their vexed identity conditions. What is to *count* as the "same effect" is the problem here, not any underlying uncertainty about what has happened.)

We can choose set X to reflect this fact (in keeping with the Guidelines): Let it contain worlds in which either (1) the bottle doesn't shatter at all, or (2) it shatters in a way very similar to the way it shatters in reality. Then for every world in X,

$$C \Rightarrow A_1$$

obtains; wherever in X the bottle shatters, we find Susie throwing her rock first. On the other hand,

$$C \Rightarrow A_2$$

may well fail in X; X can certainly contain worlds where the bottle shatters but Billy refrains from throwing. In short, A_1 is "more necessary" than A_2, provided that we choose X right. The vagueness of X, though sometimes irksome, can also break deadlocks.

Not that deadlocks must always be breakable. We ought to look with equanimity on the prospect that sometimes circumstances will fail to pinpoint a single "real cause" of an event, no matter how hard we seek. A case in point is the classic law school riddle:

> Everybody in the French Foreign Legion outpost hates Fred and wants him dead. During the night before Fred's trek across the desert, Tom poisons the water in his canteen. Then, Dick, not knowing of Tom's intervention, pours out the (poisoned) water and replaces it with sand. Finally, Harry comes along and pokes holes in the canteen, so that the "water" will slowly run out. Later, Fred awakens and sets out on his trek, provisioned with his canteen. Too late he finds his canteen is nearly empty, but besides, what remains is sand, not water, not even poisoned water. Fred dies of thirst. Who caused his death?[3]

Many will feel a temptation to insist that there *must* be an answer to this question and others like it. It is certainly true that we can agree to legislate an answer if we feel we must, and some legislative proposals will no doubt be more attractive, more intuitive, than others, but it is not clear that there are any facts—about the way the world is, or about what we really mean, or even about what we really *ought* to mean— that would settle the issue.

---

3. A doubly elaborated version of the example due originally to McLaughlin (1925), first elaborated in Hart and Honoré (1959). The Hart and Honoré version has one less twist: "Suppose A is entering a desert. B secretly puts a fatal dose of poison in A's water keg. A takes the keg into the desert where C steals it; both A and C think it contains water. A dies of thirst. Who kills him?"

Austin's Putt

Now that we have a clearer understanding of possible worlds, we can expose three major confusions about possibility and causation that have bedeviled the quest for an account of free will. First is the fear that determinism reduces our possibilities. We can see why the claim *seems* to have merit by considering a famous example proposed many years ago by John Austin:

> Consider the case where I miss a very short putt and kick myself because I could have holed it. It is not that I should have holed it if I had tried: I did try, and missed. It is not that I should have holed it if conditions had been different: that might of course be so, but I am talking about conditions as they precisely were, and asserting that I could have holed it. There is the rub. Nor does "I can hole it this time" mean that I shall hole it this time if I try or if anything else; for I may try and miss, and yet not be convinced that I could not have done it; indeed, further experiments may confirm my belief that I could have done it that time, although I did not. (Austin 1961, p. 166)

Austin didn't hole the putt. Could he have, if determinism is true? The possible-worlds interpretation exposes the misstep in Austin's thinking. First, suppose that determinism holds, and that Austin misses, and let H be the sentence "Austin holes the putt." We now need to choose the set X of relevant possible worlds that we need to canvass to see whether he could have made it. Suppose X is chosen to be the set of physically possible worlds that are *identical* to the actual world at some time t_0 prior to the putt. Since determinism says that there is at any instant exactly one physically possible future, this set of worlds has just one member, the actual world, the world in which Austin misses. So, choosing set X in this way, we get the result that H does not hold for any world in X. So it was not possible, on this reading, for Austin to hole the putt.

Of course, this method of choosing X (call it the *narrow method*) is only one among many. Suppose we were to admit into X worlds that differ in a few imperceptibly microscopic ways from actuality at t_0; we might well find that we've now included worlds in which Austin holes the putt, even when determinism obtains. This is, after all, what recent

narrow method — under determinism
should you think in narrow method

work on chaos has shown: Many phenomena of interest to us can change radically if one minutely alters the initial conditions. So the question is: When people contend that events are possible, are they really thinking in terms of the narrow method?

Suppose that Austin is an utterly incompetent golfer, and his partner in today's foursome is inclined to deny that he could have made the putt. If we let X range too widely, we may include worlds in which Austin, thanks to years of expensive lessons, winds up a championship player who holes the putt easily. That is not what Austin is claiming, presumably. Austin seems to endorse the narrow method of choosing X when he insists that he is "talking about conditions as they precisely were." Yet in the next sentence he seems to rescind this endorsement, observing that "further experiments may confirm my belief that I could have done it that time, although I did not." What further experiments might indeed confirm Austin's belief that he could have done it? Experiments on the putting green? Would his belief be shored up by his setting up and sinking near-duplicates of that short putt ten times in a row? If this is the sort of experiment he has in mind, then he is *not* as interested as he claims he is in conditions as they precisely were. To see this, suppose instead that Austin's "further experiments" consisted in taking out a box of matches and lighting ten in a row. "See," he says, "I could have made *that very putt.*" We would rightly object that his experiments had absolutely no bearing on his claim. Sinking ten short putts would have no more bearing on his claim, understood in the narrow sense as a claim about "conditions as they precisely were." We suggest that Austin would be content to consider "Austin holes the putt" possible if, in situations *very similar* to the actual occasion in question, he holes the putt. We think that this is what he meant, and that he would be right to think about his putt this way. This is the familiar, reasonable, useful way to conduct "further experiments" whenever we are interested in understanding the causation involved in a phenomenon of interest. We vary the initial conditions slightly (and often systematically) to see what changes and what stays the same. This is the way to gather *useful* information from the world to guide our further campaigns of avoidance and enhancement.

Curiously, this very point was made, at least obliquely, by G. E. Moore in the work Austin was criticizing in the passage quoted.

Moore's examples were simple: Cats can climb trees and dogs can't, and a steamship that is *now* traveling at 25 knots can, of course, also steam at 20 knots (but not, of course, in *precisely* the circumstances it is now in, with the engine set at Full Speed Ahead). The sense of "can" invoked in these uncontroversial claims, the sense called "can (general)" by Honoré (1964) in an important but neglected article, is one that *requires* us to look not at "conditions as they precisely were" but at minor variations on those conditions. *minor var on cond*

So Austin equivocates when he discusses possibilities. In truth, the narrow method of choosing X does not have the significance that he and many others imagine. From this it follows that the truth or falsity of determinism should not affect our belief that certain unrealized events were nevertheless "possible," in an *important everyday sense* of the word. We can bolster this last claim by paying a visit to a narrow domain in which we know with certainty that determinism reigns: the realm of chess-playing computer programs.

in everyday sense - det. shouldnt play role

A Computer Chess Marathon

Computers excellently instantiate the Laplacean, Democritean ideals of determinism. It is trivial to get a computer to execute a few trillion steps, and then place it back in *exactly* the same (digital) state it was in before, and watch it execute *exactly* the same few trillion steps again, and again, and again. The subatomic world in which computers live, and hence the subatomic parts of which they are made, may or may not be deterministic, but computers themselves are brilliantly designed to be deterministic in the face of submicroscopic noise and even quantum randomness, absorbing these fluctuations by being digital, not analog. The fundamental idea behind digitizing in order to produce determinism is that we can *create* inert historical facts by design. Forcibly sorting all the pivotal events into two categories—high versus low; ON versus OFF; 0 versus 1—guarantees that the micro-differences (between different high voltages, different flavors of being ON, different shades of 0) are ruthlessly discarded. Nothing is allowed to hinge on them, and they vanish without a trace, facts about actual historical variations that make *no difference at all* to the subsequent series of states through which the computer passes.

CONRAD: Computers are deterministic? You can get them to replay exactly the same trillion steps over and over? Gimme a break! Then why does my laptop crash every so often? Why does my word processor freeze on Tuesday when I was doing the very same thing that worked just fine on Monday?

You weren't doing the *very* same thing. It froze not because it is indeterministic, but because it was not in *exactly* the same state on Tuesday that it was in on Monday. Your laptop must have done something in the interval that raised a hidden "flag" or called up some part of the word processor that had never before been activated by you, which flipped a bit somewhere that got saved in its new position when you shut down, and now the word processor has stubbed its toe on that tiny change and crashed. And if you somehow manage to put it back in *exactly* the same Tuesday-morning state a second time, it will crash again.

CONRAD: What about the "random number generator"? I thought my computer had a built-in device for creating randomness on demand.

Every computer these days comes equipped with a built-in random number generator that can be consulted whenever needed by any program running on it. The sequence of numbers it generates isn't really random, but just pseudo-random: It is "mathematically compressible" in the sense that this infinitely long sequence can be captured in a finitely specified mechanism that will crank it out. Whenever you start the random number generator from a cold start—whenever you reboot your computer, for instance—it will always yield exactly the same sequence of digits, but a sequence that is as *apparently* patternless as if it were generated by genuinely random quantum fluctuations. (It is rather like a very long loop of videotape, recording the history of a fair roulette wheel over millions of spins. The loop always returns to "the beginning" when you start up your computer.) Sometimes this matters; computer programs that avail themselves of randomness at various "choice" points will nevertheless spin out exactly the same sequence of states if run over and over again from a cold start, and if you want to test a program for bugs, you will always test the same "random sample" of states, unless you take steps (easy enough) to jog the program to dip elsewhere, now and then, into the stream of digits for its next "random" number.

Suppose you install two different chess-playing programs on your computer, and yoke them together with a little supervisory program that pits them against each other, game after game, in a potentially endless series. Will they play the same game, over and over, until you turn off the computer? You *could* set it up like that, but then you wouldn't learn anything interesting about the two programs, A and B. Suppose A beats B in this oft-repeated game. You couldn't infer from this that A is a better program in general than B, or that A would beat B in a different game, and you wouldn't be able to learn anything from the exact repetition about the strengths and weaknesses of the two different programs. Much more informative would be setting up the tournament so that A and B play a succession of different games. This can be readily arranged. If either chess program consults the random number generator during its calculations (if, for instance, it periodically "flips a coin" to escape from cases where it has no handy reason for doing one thing versus another in the course of its heuristic search), then in the following game the state of the random number generator will have changed (unless you arrange to have it reinitialized), and hence different alternatives will be explored, in a different order, leading on occasion to different moves being "chosen." A variant game will blossom, and the third game will be different in different ways, resulting in a series in which the games, like snowflakes, are no two alike. Nevertheless, if you turned off the computer and then restarted it running the same program, exactly the same variegated series of games would spin out.

Suppose, then, we set up such a chess universe involving two programs, A and B, and study the results of a lengthy run of, say, a thousand games. We will find lots of highly reliable patterns. Suppose we find that A always beats B, in a thousand *different* games. That is a pattern that we will want to explain, and saying "Since the program is deterministic, A was *caused* always to beat B" would utterly fail to address our very reasonable curiosity. We will want to know what it is about the structure, the methods, the dispositions of A that account for its superiority at chess. A has a competence or power that B lacks, and we need to isolate this interesting factor. When we set about exploring the issue, we need to avail ourselves of a high-level perspective at which the "macroscopic" objects of chess decision-making appear: representations of chess pieces, board positions, evaluations of possible continuations, decisions about which continuations to pursue further, and so forth. Or it might be that the explanation lies at a lower level; it might turn out, for instance, that program A and program B are *identical* chess-move evaluators but program A is more efficiently coded so that it can explore further than program B can in the same number of machine cycles. In effect, A "thinks the same thoughts" about chess as B but just thinks faster.

It would actually be more interesting if one program didn't always win. Suppose A *almost* always beats B, and suppose A evaluates moves using a different set of principles. Then we would have something more interesting to explain. To investigate *this* causal question, we would need to study the history of the thousand different games looking for further patterns. We would be sure to find plenty of them. Some of them would be endemic to chess wherever it is played (e.g., the near certainty of B's loss in any game where B falls a rook behind), and some of them would be peculiar to A and B as particular chess players (e.g., B's penchant for getting its queen out early). We would find the standard patterns of chess strategy, such as the fact that when B's time is running out, B searches less deeply in the remaining nodes of the game tree than it does when in the same local position with more time remaining. In short, we would find a cornucopia of *explanatory* regularities, some exceptionless (in our run of a thousand games) and others statistical.

These macroscopic patterns are salient moments in the unfolding of a deterministic pageant that, looked at from the perspective of micro-causation, is pretty much all the same. What from one vantage point appear to us to be two chess programs in suspenseful combat can be seen through the "microscope" (as we watch instructions and data streaming through the computer's CPU) to be a single deterministic automaton unfolding in the only way it can, its jumps already predictable by examining the precise state of the pseudo-random number generator. There are no "real" forks or branches in its future; all the "choices" made by A and B are already determined. Nothing, it seems, is really *possible* in this world other than what actually happens. Suppose, for instance, that an ominous mating-net looms over B at time *t* but collapses when A runs out of time and terminates its search for the key move one pulse too soon. That mating-net *was never going to happen*. (This is something we could prove, if we doubted it, by running exactly the same

even though def. doesn't rob them of diff purs. [handwritten annotation]

tournament another day. At the same moment in the series, A would run out of time again and terminate its search at exactly the same point.)

So what are we to say? Is this toy world really a world without prevention, without offense and defense, without lost opportunities, without the thrust and parry of genuine agency, without genuine possibilities? Admittedly, our chess programs, like insects or fish, are much too simple agents to be plausible candidates for morally significant free will, but the determinism of their world does not rob them of their different powers, their different abilities to avail themselves of the opportunities presented. If we want to understand what is happening in that world, we may, indeed must, talk about how their informed choices cause their circumstances to change, and about what they *can* and *cannot* do. If we want to uncover the *causal regularities* that account for the patterns we discover in those thousand games, we have to take seriously the perspective that describes the world as containing two agents, A and B, trying to beat each other in chess.

Suppose we rig the tournament program so that whenever A wins a bell rings and whenever B wins a buzzer sounds. We start the marathon running, and an observer who knows nothing about the program running notes that the bell rings quite frequently, the buzzer hardly ever. What explains this regularity, she wants to know. The regularity with which A beats B can be discerned and described independently of adopting the intentional stance, and it stands in need of explanation. The only explanation—the right explanation—may be that A generates better "beliefs" about what B will do if . . . than B generates about what A will do if. . . . In such a case, adopting the intentional stance is *required* for finding the explanation.

Suppose we find two games in the series in which the first twelve moves are the same, but with A playing White in the first game and Black in the second. At move 13 in the first game, B "blunders," and it's all downhill from there. At move 13 in the second game, A, in contrast, finds the saving move, castling, and goes on to win. "B *could have castled* at that point in the first game," says an onlooker, echoing Austin. True or false? The move, castling, was just as legal the first time, so in *that* sense, it was among the "options" available to B. Suppose we find, moreover, that castling was not only one of the represented candidate moves for B, but that B, in fact, undertook a perfunctory exploration of the consequences of castling, abandoned, alas, before its

look @ similar cases

virtues were revealed. So *could* B have castled? What are we trying to find out? Looking at *precisely* the same case, again and again, is utterly uninformative, but looking at <u>similar</u> cases is, in fact, diagnostic. If we find that in many similar circumstances in other games, B *does* pursue the evaluation slightly further, discovering the virtues of such moves and making them—if we find, in the minimal case, that flipping a single bit in the random number generator would result in B's castling— then we support ("with further experiments") the observer's conviction that B could have castled then. We would say, in fact, that B's failure to castle was a fluke, bad luck with the random number generator. If, on the contrary, we find that discovering the reasons for castling requires far too much analysis for B to execute in the time available (although A, being a stronger player, is up to the task), then we will have grounds for concluding that no, B, unlike A, could <u>not</u> have castled. Castling, we may discover, was one of those moves that gets followed by "(!)" in the chess column in the newspaper, a "deep" move that was out of B's league. To imagine B castling would require too many alterations of reality; we would be committing the error mentioned earlier of making set X too large.

can't choose narrow X

In sum, using the narrow method to choose X is useless if we want to explain the patterns that are manifest in the unfolding data. It is only if we "wiggle the events" (as David Lewis has said), looking *not* at "conditions as they precisely were" but at nearby neighboring worlds, that we achieve any understanding at all. Once we expand X a little, we discover that B has additional options, in a sense both informative and morally relevant (when we address worlds beyond the chessboard). Many philosophers have assumed without specific argument that when we ask a question about what was possible, we are—and should be— interested in knowing whether, in *exactly* the same circumstances, the same event would recur. We have argued that in spite of its traditional endorsement by philosophers, this policy is *never* followed by serious investigators of possibility and is, in any event, unmotivated: It *couldn't* give you an answer that could satisfy your curiosity. The burden now rests with those who think otherwise to explain why "real" possibility demands a narrow choice of X—or why we should be interested in such a concept of possibility, regardless of its "reality."

So deterministic worlds can quite comfortably support *possibilities* of the broader, more interesting variety. Indeed, introducing *inde-*

indet. chgs nothing

terminism adds nothing in the way of worthwhile possibilities, opportunities, or competences to a universe. If in our deterministic chess tournament, program A always beats program B, then replacing the pseudo-random number generator with a genuinely indeterministic device will not help B at all. A will *still* win every time. A superior algorithm like A's will hardly stumble when faced with so inconsequential, indeed practically invisible, a change. Though pseudo-random generators may not produce genuinely random output, they come so close that for almost any purpose it makes no difference. There is one context in which it does make a practical difference: cryptography. The particular flavors of patternlessness of particular pseudo-random number-generating algorithms can eventually be sniffed out by supercomputers, putting a premium on using genuinely random numbers in these specialized contexts.[4] But aside from a context in which you have to worry about an opponent having access to your particular brand of pseudo-random number generator and using it to "read your mind," you have nothing to gain from going genuinely indeterministic. To put it graphically, the universe could be deterministic on even days of the month and indeterministic on odd days, and we'd never notice a difference in human opportunities or powers; there would be just as many triumphs—and just as many lamentable lapses—on October 4 as on October 3 or October 5. (If your horoscope advised you to postpone any morally serious decision to an odd-numbered day, you would have no more reason to follow this advice than if it told you to wait for a waning moon.)

Events without Causes in a Deterministic Universe

The vast causal independence of contemporary occasions is the preservative of the elbow-room within the Universe.

—Alfred North Whitehead, *Adventures of Ideas*

Determinism is a doctrine about sufficiency: If S_0 is a (mind-bogglingly complex) sentence that specifies in complete detail the state descrip-

---

4. If you need them, you can get sequences of truly random digits on the Web from several sources, such as www.random.org and www.fourmilab.ch/hotbits.

tion of the universe at t_0, and S_1 similarly specifies the state description of the universe at a later time t_1, then determinism dictates that S_0 is sufficient for S_1 in all physically possible worlds. But determinism tells us nothing about what earlier conditions are *necessary* to produce S_1 or any other sentence for that matter. Hence, since causation generally presupposes necessity, the truth of determinism would have little, if any, bearing on the validity of our causal judgments.

For example: According to determinism, the precise condition of the universe one second after the Big Bang (call the corresponding sentence S_0) causally sufficed to produce the assassination of John F. Kennedy in 1963 (sentence C). Yet there is no reason at all to claim that S_0 caused C. Though sufficient, we have no reason to believe that S_0 is necessary. For all we know, Kennedy might well have been assassinated anyway, even if some different conditions had obtained back during the universe's birth. How could we ever tell? We can imagine the investigation, even if we can't conduct it: Imagine that we take a snapshot of the universe at the moment of Kennedy's assassination, then alter the picture in some trivial way (by moving Kennedy 1 mm to the left, say). Sentence C, "John F. Kennedy was assassinated in 1963 (in Dealey Plaza, while riding in a motorcade . . .)," is still true, but with a microscopic difference in the atomic conditions that make it true. Then, starting from our subtly revised state description of 1963, and following the (deterministic) laws of physics in reverse, we generate a movie running all the way back to the Big Bang, obtaining a world in which S_0 subtly fails. There are highly similar possible worlds in which Kennedy is killed but S_0 is *not* the case, so the state of the universe described by S_0 is *not* the cause of Kennedy's assassination. More plausible causes of that event would include: "A bullet followed a course directed at Kennedy's body"; "Lee Harvey Oswald pulled the trigger on his gun." Conspicuously absent from this list are microscopically detailed descriptions of the universe billions of years prior to the incident. Philosophers who assert that under determinism S_0 "causes" or "explains" C miss the main point of causal inquiry, and this is the second major error.

In fact, determinism is perfectly compatible with the notion that some events have no cause at all. Consider the sentence "The devaluation of the rupiah caused the Dow Jones average to fall." We rightly treat such a declaration with suspicion; are we really so sure that among nearby

so might not be necessary

determ—some have no causes

universes the Dow Jones fell *only* in those where the rupiah fell first? Do we even imagine that every universe where the rupiah fell experienced a stock market sell-off? Might there not have been a confluence of dozens of factors that jointly sufficed to send the market tumbling but none of which by itself was essential? On some days, perhaps, Wall Street's behavior has a ready explanation; yet at least as often we suspect that no particular cause is at work.

A coin flip with a fair coin is a familiar example of an event yielding a result (heads, say) that properly *has no cause*. It has no cause because no matter how we choose the set X (ignoring Austin's mistaken advice that we consider circumstances as they *precisely* were), we will find no feature C that is necessary for heads or necessary for tails. Have you ever wondered about the apparent contradiction involved in using a coin flip as a generator of a random event? Surely the result of a coin flip is the *deterministic* outcome of the total sum of forces acting on the coin: the speed and direction of the release that imparts the spin, the density and humidity of the air, the effect of gravity, the distance to the ground, the temperature, the rotation of the earth, the distance to Mars and Venus at that time, and so forth. Yes, but this total sum has no predictive patterns in it. That is the point of a randomizing device like a coin flip, to make the result uncontrollable by making it sensitive to so many variables that no feasible, finite list of conditions can be singled out as the cause. That is why we require the coin to be flipped high, with a vigorous spin, and not just dropped from the fingers an inch above the table: We set in motion a sequence that practically guarantees that nothing will be the cause of its landing heads or tails. Notice how the strategy of flipping a coin exploits digitizing to guarantee that its outcome is causeless (if done fairly). It accomplishes just the opposite of digitizing in computers: Instead of absorbing all the micro-variation in the universe, it amplifies it, guaranteeing that the unimaginably large sum of forces acting at the moment will tip the digitizer into one of two states, heads or tails, but with no salient necessary conditions for either state.

The practice of "wiggling events" in controlled experiments is one of the great innovations of modern science, and as Judea Pearl points out, it depends on using something like coin flips to *break* the causal links that otherwise might exist between the events we wish to analyze:

Assume we wish to study the effect of some drug treatment on recovery of patients suffering from a given disorder. . . . Under uncontrolled conditions, the choice of treatment is up to the patients and may depend on the patients' socioeconomic backgrounds. This creates a problem, because we can't tell if changes in recovery rates are due to treatment or to those background factors. What we wish to do is compare patients of like backgrounds, and that is precisely what [Sir Ronald] Fisher's *randomized experiment* accomplishes. How? It actually consists of two parts, randomization and *intervention*.

Intervention means that we change the natural behavior of the individual: we separate subjects into two groups, called treatment and control, and we convince the subjects to obey the experimental policy. We assign treatment to some patients who, under normal circumstances, will not seek treatment, and we give placebo to patients who otherwise would receive treatment. That, in our new vocabulary, means *surgery*—we are severing one functional link and replacing it with another. Fisher's great insight was that connecting the new link to a random coin flip *guarantees* that the link we wish to break is actually broken. The reason is that a random coin is assumed to be unaffected by anything we can measure on a macroscopic level—including, of course, a patient's socioeconomic background. (Pearl 2000, p. 348)

Our practice in such cases belies a background assumption that seems to be widely adopted (but seldom, if ever, examined): the assumption that the only way for an event not to have a cause is for it to be strictly undetermined, to have *no* sufficient condition, no matter how diffuse and complex and uninteresting. This can lead to serious distortion of one's scientific agenda: What was the cause of World War I? Surely if we are going to be good scientific explainers, we need to find the cause! Declaring that World War I had *no* cause would be tantamount, would it not, to declaring it either a violation of the laws of nature—some miracle!—or (quantum physics to the rescue) the result of indeterministic quantum processes? No, it would not. It *could* be that no matter how historians "wiggle the events" looking for necessary antecedents for World War I in nearby possible worlds, they find that those universes in which World War I occurs do not share any common, necessary antecedent. Suppose, for instance, that in universe A, Archduke Ferdinand is assassinated and World War I subsequently breaks out. Is the former then the cause of the latter (as some of us "learned" in school)? Maybe not; perhaps in universe B, Archduke Ferdinand survives, but World War I happens anyway. And similarly, for any "cause" that historian X proposes, historian Y may be able to dream up a world in which World War I occurs without the candidate cause occurring first. The war could have been a fluke, and then persisting in arguments about "the cause" would be not just futile, but almost guaranteed to generate artifactual myths about covert causation worth pursuing further. The search for such necessary conditions is always rational, so long as we remind ourselves that there *may* be nothing to find in any particular case.[5]

One might wonder, then, why it is that causal necessity matters to us as much as it does. Let us return for a moment to chess programs A and B. Suppose our attention is drawn to a rare game in which B wins, and we want to know "the cause" of this striking victory. The trivial claim that B's win was "caused" by the initial state of the computer would be totally uninformative. Of course, the total state of the toy universe at prior moments was *sufficient* for the occurrence of the win; we want to know which features were *necessary,* and thereby understand what such rare events have in common. We want to discover those features, the absence of which would most directly be followed by B's loss, the default outcome. Perhaps we will find a heretofore unsuspected flaw in A's control structure, a bug that has only just now surfaced. Or we might find an idiosyncratic island of brilliance in B's competence, which once diagnosed would enable us to say just what circumstances in the future might permit another such victory for B. Or perhaps the victory is a huge coincidence of conditions that should provoke no repair, since the probability of their recurrence is effectively zero. This last possibility, that in the relevant sense there simply was no cause of B's victory—it was a fluke—is easy enough to understand in such a simplified context, but hard to countenance, it seems, in real-world cases.

---

5. The bias in favor of not just looking for but finding a cause is not idle, as Matt Ridley notes in his discussion of Creutzfeldt-Jakob disease, for which no cause has yet been found: "This offends our natural determinism, in which diseases must have causes. Perhaps CJD just happens spontaneously at the rate of about one case per million per year" (Matt Ridley 1999, p. 285).

Rationality *requires* that we evaluate necessary conditions at least as carefully as sufficient conditions. Consider a man falling down an elevator shaft. Although he doesn't know exactly which possible world he in fact occupies, he does know one thing: He is in a set of worlds *all* of which have him landing shortly at the bottom of the shaft. Gravity will see to that. Landing is, then, *inevitable* because it happens in every world consistent with what he knows. But perhaps *dying* is not inevitable. Perhaps in some of the worlds in which he lands, he survives. Those worlds do not include any in which he lands headfirst or spread-eagled, say, but there may be worlds in which he lands in a toes-first crouch and lives. There is some elbow room. He can rationally plan action on the assumption that living is possible, and even if he cannot *discover* sufficient conditions to guarantee survival, he may at least improve the odds by taking whatever actions are necessary, and thereby, with some luck, find himself in one of the Vastly many possible worlds in which he lives.

> **CONRAD:** Once again, what sense can this talk of *improving* his odds make? We're presupposing determinism here. He can't *change worlds.* He's in the world he is in, the actual world, and in that world he either lives or dies, and that's the end of it!

But that is true independently of determinism, and is irrelevant to the issue of the rationality of his action. Pretend we temporarily suspend this man in his plummet and allow him to peruse the Vast corner of the Library of Babel that contains biographies of somebody of his name, with his features and characteristics and history to date—the tale of a man who accidentally falls down an elevator shaft and finds himself confronting an unimaginably huge collection of books, each purporting to be his true life history. In some of these he lives and in some of these he dies (and, this being the Library of Babel, in some of these he turns into a golden teacup and is thrown at Cleopatra by a giant snail). The trouble is that although he can rule out the fantastic books on the basis of his general knowledge of how the world works, he could have no way of telling which particular book among those that have him living or dying after his fall is the truth. And assuming that determinism is true, *or false,* will not help him find the needle in this haystack. His best strategy, faced with his ineliminable uncertainty about which book tells the truth, is to look for general patterns of predictive saliency—causes and effects—and be guided by the anticipations these commend to him. But how is he to do that? Not a problem: He is already designed to be caused to do that, by eons of evolution. If he didn't have these talents, he wouldn't be here. He is the product of a design process that has created species of anticipator-avoiders to whom this trick is second nature. They are not perfect, but they do much better than chance. Compare, for instance, the prospects of beings who are confronted by the opportunity to win a million dollars by calling a coin toss or by rolling two dice and getting snake eyes. Some of them reason fatalistically: "It makes no difference which method I choose; the odds of my throwing snake eyes are either 0 or 1. I don't know which fate is already determined, and the same is true for my calling the coin." Others act on the conviction that the 1–in–2 chance of calling the coin is much better than the 1–in–36 chance of rolling snake eyes, and opt for the coin toss. Not surprisingly, people so designed have outperformed the fatalists, who can be seen from the perspective of history to have a design flaw.

Will the Future Be Like the Past?

And now, at last, we are ready to confront the third major error in thinking about determinism. Some thinkers have suggested that the truth of determinism might imply one or more of the following disheartening claims: All trends are permanent, character is by and large immutable, and it is unlikely that one will change one's ways, one's fortunes, or one's basic nature in the future. Ted Honderich, for example, has maintained that determinism would somehow squelch what he calls our life-hopes:

> If things have gone well for a person, there is more to hope for in what follows on the assumption that the entire run of his or her life is fixed. . . . If things have not gone well, or not so well as was hoped, it is at least not unreasonable to have greater hopes on the assumption that the whole of one's life is not fixed, but is connected with the activity of the self. . . . Given the sanguine

premise of our reasonableness, there is reason to think that we do
not tend to the idea of a fixed personal future. (Honderich 1988,
pp. 388–89)

Clearly such anxieties originate in a vague sense that true possibilities
(for an improved lot, say) disappear under determinism. But this is a
mistake. The distinction between being a thing with an open future
and being a thing with a closed future is strictly independent of determinism. In general, there is no paradox in the observation that certain
phenomena are *determined* to be changeable, chaotic, and unpredictable,
an obvious and important fact that philosophers have curiously
ignored. Honderich finds disturbing the notion that we might have a
"fixed personal *future*," but the implications of this notion are entirely
distinct from the implications of having a "fixed personal *nature*." It
could very well be one's "fixed"—that is, determined—personal *future*
to be blessed with a protean *nature,* highly responsive to the "activity
of the self." The total set of personal futures, "fixed" or not, contains
all sorts of agreeable scenarios, including victories over adversity, sub-
jugations of weakness, reformations of character, even changes of luck.
It could be just as determined a fact that you *can* teach an old dog new
tricks as that you can't. The question to ask is: Are old dogs the kinds
of things that can be taught new tricks? If they aren't, we don't want
to be like old dogs. We rightly care about being the sorts of entities
whose future trajectories are not certain to repeat the patterns found
in the past, and the general thesis of determinism has no implications
at all about such issues.

Consider the simple deterministic Life worlds. At one level
nothing ever changes; pixels do the same thing over and over forever,
following the simple rule of physics. At another level, we see different
kinds of worlds. Some worlds are just as changeless from a bird's-eye
view as they are at the atomic level, a field of still lifes and flashers, say,
flashing for eternity. No drama, no suspense. Other worlds "evolve"
continuously, never returning to the same state twice, in either a pat-
terned way, growing predictably, creating a steady stream of identical,
equally spaced gliders, for instance, or in an apparently patternless way,
with myriad growing, shifting, colliding swarms of pixels. In these
worlds is the future like the past? Yes and no. The physics is eternally
changeless, so the micro-events are always the same. But at a higher

[handwritten: natures can chg under det.]

level, the future may be variegated: It may contain some patterns that are like the patterns of its past, and it may contain others that are entirely novel. In some deterministic worlds, that is, there are things whose *natures* change over time, so determinism does not imply a fixed nature. A small, but heartening, fact. There are more to come.

Some Life worlds contain competitions, and even though Laplace's demon knows exactly how each competition will end, there may be genuine drama and suspense for lesser intelligences, who cannot know, from their limited perspective, how the contest will end. Consider, for instance, those Life worlds in which there is a Universal Turing Machine running our program in which A is playing B in chess. Chess is a game of "perfect information"; in this regard it is unlike card games, in which you keep your cards concealed from your opponent (and in which no opponent knows what card will come up next in the deck). So both A and B have common and total information about the state of the chess game in progress and the possibilities that lie ahead. They nevertheless come to have differing inventories of hard-won expectations about the probable future moves of their opponents—and themselves. The contest is to use the shared information to generate proprietary information on which to base one's choice of move, and the *explanation* of why A beats B (if it does, when it does) must be in terms of its superior capacity to generate, and use, information about the uncertain, open future (from its perspective). Every finite information-user has an epistemic horizon; it knows less than everything about the world it inhabits, and this unavoidable ignorance guarantees that it has a *subjectively* open future. Suspense is a necessary condition of life for any such agent.[6]

[handwritten margin note: nobody knows everything subj. open future]

But set aside subjective suspense, and change of nature. What about *improvement?* Can there be not just improvement, but *self-generated* improvement in a deterministic world? Can an agent in a deterministic world realistically hope to improve its lot? Once again, the answer to this question has nothing to do with determinism and

6. Laplace's demon instantiates an interesting problem first pointed out by Turing, and discussed by Ryle (1949), Popper (1951), and MacKay (1960). No information-processing system can have a complete description of itself—it's Tristram Shandy's problem of how to represent the representing of the representing of . . . the last little bits. So even Laplace's demon has an epistemic horizon and, as a result, cannot predict its own actions the way it can predict the next state of the universe (which it must be outside).

only to do w/ design   *involve chance?*

everything to do with *design*. Programmers have already demonstrated how deterministic computer algorithms can adapt themselves to changes in the environment and learn from their mistakes. We have postponed invoking a talent for learning in chess programs A and B, not wanting to distract attention from the other issues under discussion, but consider what happens when we incorporate a capacity to learn from experience in one of the contestants. If initially mediocre B possesses the capacity to learn and A does not, then we may ultimately find B emerging victorious. One of the products of B's history of competitions against A, the fruits of its very own labors, you might say, could be B's evolving a structure that gave it an improved competence, and hence an improved lot in life. B changes from a perennial loser into a regular winner. Suppose B has this sort of learning structure in a deterministic world; its enviable capacity will not improve *at all* with the introduction of a genuinely indeterministic random-number generator. Nor will adding indeterminism to the universe help open up B's future if it lacks this ability to learn.

self-improvement The conditions under which such self-improvement occurs (non-miraculously) are precisely the conditions under which something—either a hacker God, or evolution, or B's instructor, or B itself—discerns the causes responsible for victory and installs designs that enhance the likelihood of the presence of those causes at the right times in the future. There is, then, a familiar reason to design a program to learn from experience: In the future it may encounter a similar situation, and what happens then can be influenced by what it learns now. This is because what happens then will depend on what it decides then; whether or not to castle, for instance, will be *up to it* in one important sense. Whether the rules of chess remain constant will not be up to it, nor will its opponent's moves be up to it; its own moves, however, will be up to it in the sense that matters: They will be the outcome of *its* exploratory and deliberative processes.

Similarly, contrast a fish confronting a baited hook with a fish confronting a swiftly approaching net; whether the first fish takes the bait is up to the fish, but whether the second fish enters the net is probably not. Do fish have free will, then? Not in a morally important sense, but they do have control systems that make life-or-death "decisions," which is at least a necessary condition for free will. In Chapter 4 we will consider whether there is another, more weighty sense of "up to" that applies to us (if we are moral agents) but not to deterministic chess-playing computers—or fish.

We live in a world that is subjectively open. And we are *designed* by evolution to be "informavores," epistemically hungry seekers of information, in an endless quest to improve our purchase on the world, the better to make decisions about our subjectively open future. The moon is made of the same sort of stuff that we are, obeying the same laws of physics, but its nature, *unlike* ours, is fixed. Moreover, unlike us, its nature is nothing to it. It is not equipped to care for itself in the slightest. The difference between us and the moon is not a difference of physics; it is a higher-level difference of design. We are the product of a massive, competitive design process; the moon is not. This design process, natural selection, famously involves "random" mutation as its ultimate Generator Of Diversity. We have seen that computer programs—and controlled experiments more generally—make use of such generators of diversity to much the same effect: to drive exploratory processes into new patterns, and out of old patterns. But we have also seen that this welcome source of diversity need not be truly random in the sense of *indeterministic*.

To say that if determinism is true, your *future* is fixed, is to say . . . nothing interesting. To say that if determinism is true, your *nature* is fixed, is to say something false. Our natures aren't fixed because we have evolved to be entities *designed* to change their natures in response to interactions with the rest of the world. It is confusion between having a fixed *nature* and having a fixed *future* that mis-motivates the anguish over determinism. The confusion arises when one tries to maintain two perspectives on the universe at once: the "God's eye" perspective that sees past and future all laid out before it, and the engaged perspective of an agent *within* the universe. From the timeless God's-eye perspective nothing ever changes—the whole history of the universe is laid out "at once"—and even an indeterministic universe is just a static branching tree of trajectories. From the engaged agent's perspective, things change over time, and agents change to meet those changes. But of course not *all* change is possible for us. There are things we can change and things we can't change, and some of the latter are deplorable. There are many things wrong with our world, but determinism isn't one of them, even if our world is determined.

So having set aside the fear of physical determinism, we can direct our attention to the biological level at which we might actually explain how it can be that *we* are free, when other entities in our world, made of the same kind of stuff, are not free at all. And as usual, when the topic is biology, we will find that there are all manner of different kinds and grades of freedom. The freedom, such as it is, of a chess-playing computer living on the Life plane is a toy, a mere cartoon sketch of the kind of freedom we're interested in. But we *are* interested in this kind of freedom, and it helps to begin with the simplest imaginable model of it, and to confirm that it is compatible with determinism.

> **CONRAD:** OK, you have shown that Austin was wrong. But it turns out that he wasn't interested in *real* possibility at all; he was interested in his putting game! And you are right that the way to check up on that is by hitting a few putts and seeing how many go in. As you show, there is a sense of competence, of *can do,* that applies equally well to human agents and such contraptions as chess-playing computers (and can openers, for that matter). But all this shows is that answering *that* kind of question is not even addressing the question that interests me: Could Austin have made *that very* putt? And the answer to that question must be "no" in a deterministic world.

Very well, if you insist. Maybe there is a sense of "possible" in which Austin could not possibly have made that very putt, if determinism is true. Now why on earth should we care about your question? Aside from idle metaphysical curiosity, what interest should we take in whether or not Austin could have made the putt *in your sense?*

The incompatibilists do have an answer to this question, and before we can comfortably turn back to evolution, we should give them a chance to present it. The next chapter is devoted to looking at their best answer to date. Those who are already persuaded that determinism is just not the issue may pass over Chapter 4, but they will miss some incidental discoveries about the nature of our freedom that are quite independent of the quest for indeterminism that uncovered them.

Chapter 3

Our everyday thinking about possibility, necessity, and causation seems to con-flict with determinism, but this is an illusion. Determinism doesn't imply that whatever we do, we could not have done otherwise, that every event has a cause, or that our natures are fixed.

Chapter 4

A sympathetic look at an ambitious indeterministic model of decision-making exposes the motivations as well as the problems that beset any theorist that fol-lows that path. What libertarians plausibly claim to need can be provided with-out indeterminism, and indeterminism cannot make any difference that could make a moral difference.

Notes on Sources and Further Reading

Judea Pearl's *Causality: Models, Reasoning, and Inference* (2000), which I dis-covered while preparing the final draft of this book, raises questions about the Taylor/Dennett way of putting things in terms of possible worlds, while opening up tempting alternative accounts. It will be no small labor to digest these and, if need be, reformulate our conclusions, which we do not think are directly challenged. This is work for the future.

For more on possibility, see *Darwin's Dangerous Idea* (Dennett 1995), Chapter 5, "The Possible and the Actual" and, especially, "Pos-sibility Naturalized" (pp. 118–23). See also the thought experiment ("Two Black Boxes," pp. 412–22), in which it can be seen that scien-tists could have *total knowledge* of the micro-causal processes occurring (deterministically) in this phenomenon and yet be completely baffled about the macro-causal regularity they observe and wish to explain.

For more on pseudo-random numbers and their uses in control and free will, see *Elbow Room* (Dennett 1984), pp. 66–67 and elsewhere.

Published in nine volumes between 1759 and 1766, Laurence Sterne's comic novel *Tristram Shandy* purports to be an autobiography, but winds itself into recursive loops of reflection and reaction and meta-reaction, a task unfinished and unfinishable.

Chapter 4

A HEARING FOR LIBERTARIANISM

The traditional problem of free will is introduced by the proposition that *if determinism is true, then we don't have free will*. This proposition expresses *incompatibilism,* and it certainly seems plausible at the outset. Many who have thought long and hard about it still think it's true, so before returning to my project, which denies it outright, let's take it for a test drive to see what its appeal is, and what its strengths are, as well as its weaknesses.

The Appeal of Libertarianism

If we accept the proposition as it stands, two paths open up, depending on which half of the proposition we cling to:

> *Hard determinism:* Determinism is true, so we don't have free will. Hard-headed scientific types sometimes proclaim their acceptance of this position, even declaring it a no-brainer. Many of them would add: And if determinism is false, we *still* don't have free will—we don't have free will in any case; it's an incoherent concept. But they typically excuse themselves from exploring the question of how they then justify the often strongly held moral convictions that continue to guide their lives. Where does this leave us? What sense are we to make of human striving, praising, blaming? In Chapter 1 we encountered the spiral into the abyss that beckons at this juncture. Are there any stable alternatives to this threatened moral nihilism? (The hard determinists among you may find in subsequent chapters that your *considered* view is that

whereas free will—as you understand the term—truly doesn't exist, something *rather like* free will does exist, and it's just what the doctor ordered for shoring up your moral convictions, permitting you to make the distinctions you need to make. Such a soft landing for a hard determinist is perhaps only terminologically different from *compatibilism,* the view that free will and determinism are compatible after all, the view that I am defending in this book.)

Libertarianism: We do have free will, so determinism must be false; *in*determinism is true. Since, thanks to quantum physicists, the received view among scientists today is that indeterminism *is* true (at the subatomic level and, by implication, at higher levels under various specifiable conditions), this can look like a happy resolution of the problem, but there is a snag: How can the indeterminism of quantum physics be harnessed to give us a clear, coherent picture of a human agent exercising this wonderful free will?

This meaning of *libertarianism,* by the way, has nothing to do with the political sense of the term. There are probably more left-leaning than right-leaning philosophers who defend this kind of libertarianism, but only because there are probably more left-leaning philosophers in general. It might be true that political right-wingers who have thought about it tend to favor free will libertarianism, and religious conservatives are drawn to it, if only by being repelled by all the alternatives, but free will libertarians are not committed to any particular view about the powers of the state vis-à-vis the citizens. They agree that free will depends on indeterminism but they divide rather sharply on the snag just mentioned: How, exactly, could subatomic indeterminism yield free will? One group simply declares that this is somebody else's problem, a job for neuroscientists, perhaps, or physicists. All they are concerned with are what we might call the top-down constraints of moral responsibility: For a human agent to be properly held responsible for something done, it must be the case one way or another that the agent's choice of this action was not determined by the total set of physical conditions that obtained prior to the choice. "We philosophers are in charge of setting the *specs* for a free agent; we leave the problem of *implementation* of those specs to the neuro-engineers." Another, smaller group has appreciated that this division of labor is not always a good idea; the very coherence of the libertarian specs is called into question by the difficulties one encounters in trying to implement them. Moreover, it turns out that the attempt to devise a positive account of indeterministic human choice pays dividends that are independent of the assumption of indeterminism.

The best attempt so far is by Robert Kane, in his 1996 book, *The Significance of Free Will.*[1] Only a libertarian account, Kane claims, can provide the feature we—some of us, at least—yearn for, which he calls Ultimate Responsibility. Libertarianism begins with a familiar claim: If determinism is true, then every decision I make, like every breath I take, is an effect, ultimately, of chains of causes leading back into times before I was born. In the previous chapter I argued that determination is not the same as causation, that knowing that a system is deterministic tells you nothing about the interesting causation—or *lack* of causation—among the events that transpire within it; but that's a controversial conclusion, flying in the face of a long tradition. Some may view it as, at best, an eccentric recommendation about how to use the word "cause," so let's set it aside temporarily and see what happens if instead we stick with tradition and treat determinism as the thesis that each state of affairs *causes* the succeeding state. As many have claimed, then, if my decisions are caused by chains of events leading back before my birth, I can be *causally* responsible for the results of my deeds in the same way a tree limb falling in a windstorm can be causally responsible for the death of the person it falls on, but it is not the limb's *fault* that it was only as strong as it was, or that the wind blew so fiercely, or that the tree grew so close to the footpath. To be morally responsible, I have to be the ultimate source of my decision, and that can be true only if no earlier influences were *sufficient* to secure the outcome, which was "truly up to me." Harry Truman had a famous sign on his desk in the Oval Office of the White House: "The Buck Stops Here." A human mind has to be a place where the buck stops, Kane says, and only libertarianism can provide this kind of free will, the kind that can give us Ultimate Responsibility. A mind is an arena of "willings (choices, decisions, or efforts)" and

---

1. Followed up with a response to his critics in "Responsibility, Luck, and Chance: Reflections on Free Will and Indeterminism" (1999).

If these willings were in turn caused by something else, so that
the explanatory chains could be traced back further to heredity
or environment, to God, or fate, then the ultimacy would not lie
with the agents but with something else. (Kane 1996, p. 4)

Libertarians have to find a way of breaking these ominous causal chains
in the agent at the time of decision, and as Kane acknowledges, the
inventory of libertarian models so far devised is a zoo of hopeless mon-
sters. "Libertarians have invoked transempirical power centers, non-
material egos, noumenal selves, non-occurrent causes, and a litany of
other special agencies whose operations were not clearly explained" (p.
11). He sets out to correct that deficiency.

      Before turning to his attempt, however, we should note that
some libertarians don't see this as a deficiency. Unrepentant dualists and
others actually embrace the idea that it would take a miracle of sorts
for there to be free will. They are sure in their bones that free will, real
free will, is strictly impossible in a materialist, mechanist, "reductionist" world—and so much the worse for that materialist vision! Consider, for instance, the doctrine known as "agent causation." Roderick
Chisholm, the chief architect of the contemporary version of this
ancient idea, defines it thus:

If we are responsible . . . then we have a prerogative which some
would attribute only to God: each of us, when we act, is a prime
mover unmoved. In doing what we do, we cause certain events
to happen, and nothing—or no one—causes us to cause those
events to happen. (Chisholm 1964, p. 32)

How do "we" cause these events to happen? How does an *agent* cause
an effect without there being an event (in the agent, presumably) that
is the cause of that effect (and is itself the effect of an earlier cause, and
so forth)? Agent causation is a frankly mysterious doctrine, positing
something unparalleled by anything we discover in the causal processes
of chemical reactions, nuclear fission and fusion, magnetic attraction,
hurricanes, volcanos, or such biological processes as metabolism,
growth, immune reactions, and photosynthesis. Is there such a thing?
When libertarians insist that there must be, they play into the hands of
those at the other pole, the hard determinists, who are content to let
the libertarians' uncompromising definition of free will set the terms

of the debate, so that they can declare, with science as their ally, so much the worse for free will. I find that those who take it as just obvious that free will is an illusion tend to take their definition of free will from radical agent-causation types.

This polarization is probably inevitable. When the stakes are high, one should be cautious, but excess caution leads to hardened positions and paranoia about "erosion." If you're not part of the solution, you're part of the problem, as they say. Beware the thin edge of the wedge, the slippery slope. If you give them an inch, they'll take a mile. Caution can also lead to a sort of unwitting self-caricature, however. In their zeal to protect something precious, people sometimes decide to dig the moat too far out, thinking that it is safer to defend too much than risk defending too little. The result is that they end up trying to defend the indefensible, clinging to an extreme position that is actually vulnerable only because of its exaggeration. Absolutism is an occupational hazard in philosophy in any case, since radical, hard-edged positions are easier to define clearly, are more memorable, and tend to attract more attention. Nobody ever became a famous philosopher by being a champion of ecumenical hybridism. On the topic of free will this tendency is amplified and sustained by tradition itself: As philosophers for two millennia have said, either we have free will or we don't; it's all, or nothing at all. And so the various compromise proposals, the suggestions that determinism is compatible with at least *some* kinds of free will, are resisted as bad bargains, dangerous subversions of our moral foundations.

Libertarians have long insisted that the *compatibilist* sorts of free will I am describing and defending are not the real thing at all, and not even an acceptable substitute for the real thing, but rather a "wretched subterfuge," in the oft-quoted phrase of Immanuel Kant. Two can play this disparagement game. Watch. According to us compatibilists, *libertarians* seem to think that you can have free will only if you can engage in what we might call *moral levitation*. Wouldn't it be wonderful to be able to levitate—and then to dash off in any direction with the merest flick of a whim? I'd love to be able to do that, but I can't. It's impossible. There are no such miraculous things as levitators, but there are some pretty good near-levitators: Hummingbirds, helicopters, blimps, and hang gliders come to mind. Near-levitation isn't good enough, though, for libertarians, who say, in effect:

If your feet are on the ground, the decision isn't really yours—it's really planet Earth's decision. The decision isn't *made by you* but is rather a mere summation of causal trains intersecting in your body, a mobile bump on the surface of the planet, buffeted by influences, answerable to gravity. Real autonomy, real freedom, requires that the chooser be somehow suspended, isolated from the push and pull of all those causes, so that when decisions are made nothing causes them except *you!*

Those are the caricatures. They have their uses, but now let's get serious and consider Kane's intrepid attempt to fill in the gaps and provide a libertarian model of responsible decision-making. Acknowledging that "*freedom* is a term with many meanings," Kane grants that "*even if we lived in a determined world,* we could meaningfully distinguish persons who are free from such things as physical restraint, addiction or neurosis, coercion or political oppression, from persons not free from these things, and we could allow that these freedoms would be worth preferring to their opposites even in a determined world" (Kane 1996, p. 15). So some freedoms worth wanting are compatible with determinism, but "human longings transcend" those freedoms; "there is *at least one* kind of freedom that is incompatible with determinism, and it is a *significant kind of freedom worth wanting.*" It is "the power to be the ultimate creator and sustainer of one's own ends or purposes" (p. 15).

It is commonly supposed that in a deterministic world, there are no *real* options, only apparent options. In the previous two chapters, I have shown that this is an illusion, but if it is, it is also remarkably resilient and tempting. If determinism is true, then there is at any instant exactly one physically possible future, so since every choice has already been determined, all of life is just the playing out of a script that was fixed at the dawn of time. With no real options, no branch points in one's trajectory through history, it seems you can hardly be the *author* of your acts; you are more like an actor in a play, speaking your lines with apparent conviction, committing your "crimes" with grace or clumsiness, whichever has been fixed in the stage directions. Compelling, isn't it? But false. Probably the best way to drive home the surprising conclusion that this is just wrong—a panic reaction that is simply not justified by the premise of determinism—is to give the other side their best shot at saying what *would* give us real options. The challenge Kane faces is to describe a way our *apparent* decision-making could be *real* decision-making, and he wants to do this without postulating any supernatural entities or mysterious forms of agency. He is, like me, a naturalist, who assumes that we are creatures of the natural order whose mental activity is dependent on the operations of our brains. This requirement of naturalism sets some questions well worth asking. (In later chapters, we'll look more closely at what contemporary cognitive neuroscience and psychology have to say about decision-making, to see what interesting things happen when we get more ambitious and try to put in more of the details.)

Where Should We Put the Much-needed Gap?

A legendary book review begins, "This book fills a much-needed gap," and whether or not the author of that review meant what he said, Kane definitely needs a gap, a hiatus in determinism, and he wants to install it in what he calls the *faculty of practical reason* in the brain. He describes this faculty in terms of its input, its output, and what sometimes happens during the process that takes it from input to output (see Figure 4.1). These three phenomena are distinguished by Kane in terms of three senses of *will:*

> (i) *desiderative* or *appetitive will:* what I *want, desire,* or *prefer* to do
>
> (ii) *rational will:* what I *choose, decide,* or *intend* to do
>
> (iii) *striving will:* what I *try, endeavor,* or make an *effort* to do. (Kane 1996, p. 26)

Roughly, will of type (i) provides the input to the faculty of practical reason, which yields type (ii) will as output when all goes well. When there is a strain on the machinery we get (iii), which always implies a resistance, generating striving or heightened effort. This all sounds quite familiar and right. When we are undecided, we stoke up our minds with whatever relevant preferences or desires occur to us (i), remind ourselves of relevant facts or beliefs, and then mull. Our mullings, easy or effortful (iii), eventually terminate in decisions (ii). "If there is indeterminacy in free will, on my view, it must come somewhere between the input and the output" (Kane 1996, p. 27).

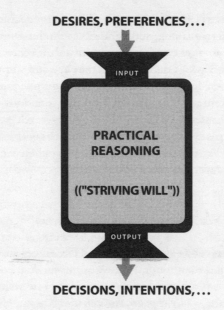

DESIRES, PREFERENCES, ...

INPUT

**PRACTICAL
REASONING**

(("STRIVING WILL"))

OUTPUT

DECISIONS, INTENTIONS, ...

Figure 4.1   Faculty of Practical Reasoning

Kane sets up an example so we can see such a system in action: Consider the case of a businesswoman "who is on the way to a meeting important to her career when she observes an assault in an alley. An inner struggle ensues between her moral conscience, to stop and call for help, and her career ambitions, which tell her she cannot miss this meeting" (Kane 1996, p. 126). He ventures the idea that this struggle might set up two "recurrent and connected neural networks"—one for each side of the issue. These two interconnected networks feed back on each other, interacting in multifarious ways, interfering with each other, and generally churning along until one of them wins the tug-of-war, at which time the system settles, outputting a decision.

Such networks circulate impulses and information in feedback loops and generally play a role in complex cognitive processing in the brain of the kind that one would expect to be involved in human deliberation. Moreover, recurrent networks are nonlinear, thus allowing (as some recent research suggests) for the possibility of *chaotic activity* [my italics—DCD], which would contribute to the plasticity and flexibility human brains display in creative problem solving (of which practical deliberation is an example). The input of one of these recurrent networks consists of the woman's moral motives, and its output the choice to go back; the input of the other, her career ambitions, and its output, the choice to go on to her meeting. The two networks are connected, so that the *indeterminism that made it uncertain* [my italics—DCD] that she would do the moral thing was coming from her desire to do the opposite, and vice versa—the indeterminism thus arising, as we said, from a conflict in the will. (Kane 1999, pp. 225–26)

In term of netuks not enough for indet.

Before we go any further, we need to separate two issues that are run together in this passage. The "chaotic activity" Kane mentions here is *deterministic* chaos, the *practical* unpredictability of certain sorts of phenomena that are describable in plain old Newtonian physics. As Kane recognizes, two networks interacting chaotically would not in themselves create any indeterminism, so if there is any "indeterminism that made it uncertain," it has to come from elsewhere. This is a key point. Kane is not alone in seeing the importance of chaos in decision-making, but it is his idea to *supplement* chaos with a smidgen of quantum randomness, following, with many others, in the wake of Roger Penrose (1989, 1994). The question we need to consider is whether any important work is being done by Kane's extra ingredient, and for this we need to get clearer about what a chaotic phenomenon is.

Consider the Hyatt New Departure Ball Bearing exhibit. For many years, the Museum of Science and Technology in Chicago displayed a glass case in which an astonishing phenomenon unfolded, hour after hour. This exhibit, donated by a branch of General Motors, showed an endless parade of little steel balls rolling out of a little hole in the back of the exhibit, falling several feet onto the highly polished top of a beautifully machined cylindrical steel "anvil," bouncing high in the air through a ring that rotated like a coin spun on a tabletop (so that timing the leaps through the rotating ring had to be exquisitely precise), and then bouncing off a second anvil up to a small hole in the back of the case, through which they all made their precise exits: Bounce, bounce, swish, bounce, bounce, swish, hundreds of times an hour. The sign on it said: "This machine demonstrates the accuracy of manufacture and uniformity of physical properties of the balls used in ball bearings." Once the two anvils were properly adjusted, it would run for days on end, with each ball following exactly the trajectory of its predecessor, a perfectly predictable, reliable, deterministic unfolding, a powerful demonstration that physical properties can fix one's destiny—at least if one is a little steel ball. Its predictability could have been shattered, however, by simply doubling the number of anvils (so that each ball had to take four bounces before exiting) and turning the anvils on their sides, so that the balls had to bounce off the rounded walls of the cylinders instead of their ultra-flat tops. The margins of error for machining the balls and adjusting the anvils would shrink vanishingly close to zero.[2] The mere presence of onlookers on the other side of the glass would create enough variable *gravitational* interference to upset the most exacting of calculations and cause many of the balls to miss their final destinations!

This kind of chaos is deterministic, but not for that reason uninteresting; it could indeed, as Kane says, "contribute to the plasticity and flexibility human brains display." In recent years the powers of such chaos, and "non-linearity" more generally, have been explored and amply demonstrated in many models alluded to by Kane. Some of this research has been heralded by critics as the death knell of Artificial Intelligence or, more specifically, the symbol-crunching variety known as GOFAI—Good Old Fashioned Artificial Intelligence (Haugeland 1985), and the impression has been created in many quarters that non-linear neural networks have wondrous powers altogether off-limits to mere computers, with their clunky, brittle algorithmic programs. But what many fans of neural networks have overlooked is the fact that the very models they advertise to prove their point are *computer* models, not just strictly deterministic but even, down in the engine room, algorithmic. They are non-algorithmic only at the highest level. (Can a whole be "freer" than its parts? Here is one way it can.) Even such an astute commentator as Paul Churchland can fall into this tempting trap. Correctly disparaging Roger Penrose's attempt to enlist quantum physics against the dread algorithms of AI, Churchland writes:

---

2. The physicist Michael Berry (1978) has done the calculations for predicting the trajectory of steel balls off the round posts in pinball machines. Three rebounds takes us beyond the limits of feasible calculation.

> *One need not look so far afield as the quantum realm to find a rich domain of nonalgorithmic processes.* The processes taking place within a *hardware* [my italics—DCD] neural network are typically nonalgorithmic, and they constitute the bulk of the computational activity going on inside our heads. They are nonalgorithmic in the blunt sense that they do not consist in a series of discrete physical states serially traversed under the instructions of a stored set of symbol-manipulating rules. (Paul Churchland 1995, pp. 247–48)

Notice the insertion of the word "hardware" here. Without it, what Churchland says would be false. In fact, all the results he discusses (NETTalk, Elman's grammar-learning networks, Cottrell and Metcalfe's EMPATH, and others) were produced not by "hardware neural networks" but by virtual neural networks simulated on standard computers. And so, at a low level, every one of these demonstrations *did* "consist in a series of discrete physical states serially traversed under the instructions of a stored set of symbol-manipulation rules." This is not the level at which to explain their power, of course, but it is an algorithmic level. Nothing these programs do transcends the limits of Turing computability. Just as we had to go to the chess-playing level to explain the difference in powers between programs A and B in Chapter 3, we have to go to the neural-network-modeling level to explain the remarkable powers of these simulated networks, but in both cases what is going on at the micro-level is a deterministic, digital, algorithmic process. The very models Churchland discusses so favorably are implemented as computer programs—algorithms, from the point of view of the limits of computability. So, unless he wants to disavow his own favorite examples, he must grant, after all, that algorithmic processes can exhibit the powers he thinks are crucial to the explanation of mentality. But then his claim that *hardware* neural networks are nonalgorithmic, even if true, would not play any role in explaining the powers they exhibit—since algorithmic approximations thereof have all the necessary powers.[3]

The simple Life world agents considered in Chapter 2 and the computer chess programs considered in Chapter 3 were both digital and deterministic, and so, for all their extra powers, are computer sim-

---

3. This paragraph is drawn, with revisions, from Densmore and Dennett 1999.

ulations of non-linear neural networks. Churchland's extra ingredient—hardware in place of virtual machine software—adds nothing to the powers of neural networks. Or if it does, nobody has given us any reason to think so.[4] Does Kane's extra ingredient—quantum level indeterminism—do any more work? To answer this question, we need to consider the details. Where and how should Kane insert the indeterminism he wants?

Kane's Model of Indeterministic Decision-making

What should the faculty of practical reasoning do, and how should it do it? What are the specs, as an engineer would say, of this deciding-device? Kane tells us that it should somehow discern the weight of the various reasons and preferences fed to it, and tip the scales in favor of the reason the agent "wants to act on more than he or she wants to act on any other reasons (for doing otherwise)." He adds the further proviso that felicitous or successful cases of the faculty in action should not be the result of coercion or compulsion (Kane 1996, p. 30). Kane deliberately leaves open at the outset the question of whether the faculty operates deterministically, since he wants to argue that, for libertarian free will to emerge from the faculty, this extra feature of indeterminism must be installed. In considering the specs for a faculty of practical reason, it helps to go beyond Kane's minimal conditions and consider some of the sorts of *incompetence* you wouldn't want your faculty to exhibit.

(1) It gives no output at all—it's just broken. You are unable to think about what to do next.

---

4. There *might* be a reason, implicit in my discussion of the role of collision in creativity in Chapter 2. It might be that no feasible computer simulation, no virtual world small enough to simulate, can have the mixture of noisiness and quietness required for open-ended creative power. That would not be germane to Churchland's claim about neural networks, but it might be true. The work of Adrian Thompson (e.g., Thompson et al. 1999) on evolutionary electronics suggests from a different quarter that software cannot always substitute for hardware in the exploration of design space. Thompson has created hardware chips with abilities that do not depend on their software-handling capabilities but rely instead on undesigned interactions at the microphysical level that can be selected for by artificial evolution.

(2) It has too narrow a bandwidth (it can't handle simultaneously all your wants or desires or preferences, and thrashes away, unable to digest its huge input).

(3) It gives output too slowly for the world you live in.

(4) It has Hamlet's problem (infinite loop) and delays its output indefinitely.

(5) It fails for particular *sorts* of input (advice from Mom, considerations of patriotism, sex, or tenure . . .).

(6) It gives the *wrong* output for the input (e.g., you definitely *prefer* human rights to having an ice cream at time *t*, but your faculty has you *decide* to buy an ice cream instead of putting the money in the Amnesty International box).

This last suggestion raises an interesting question about weakness of will, and the striving will—Kane's type (iii)—that arises when there is resistance and something has to give. Where is the *clutch* on this mechanism? Is it outside the faculty or inside?

The example given in (6) puts the clutch inside the faculty, allowing unwanted slippage *between* the input and the output: You arrive at an unwanted decision. But apparently there's another sort of case: Your practical reasoning works just fine so that you do *decide* to spend the money on human rights, but (darn it) the clutch slips *after* you make the decision and you end up buying the ice cream instead of doing *what you decided to do*. (See Figure 4.2.) Are these really two different cases? If so, what is the difference, and why is it important? When is a decision really a decision? This is not the only problem about the boundaries that we will encounter.

What if your faculty of practical reasoning were to give different outputs for the very same inputs? Would this be a flaw? Usually we want systems to be reliable, and by this we mean that we count on them always to give the same output—the *best* output, whatever it is—for each possible input. Consider your hand calculator as an example. Sometimes, however, when the best output is not definable or we specifically want the system to introduce "random" variation into the surrounding supersystem, we are content to have it give different outputs for the very same input. The standard way to achieve this is to

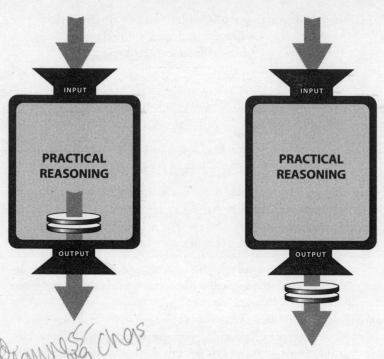

Figure 4.2   Clutch Positions, Inside and Out

incorporate a pseudo-random number generator in the system, serving the function of a coin flip (by generating either a 0 or a 1 every
time it is asked) or the throw of an ordinary six-sided die (by generating a number between 1 and 6 every time it is asked) or the spin of a
wheel of fortune (by generating a number between 1 and *n* every time
it is asked). Kane wants something better than pseudo-randomness.
He wants genuine randomness, and he proposes to get it by supposing
there is some kind of quantum-fluctuation amplifier in the neurons.
As we saw in the previous chapter, this wouldn't make his model any
more flexible or open-ended, more capable of improving itself or
learning. It wouldn't give his system any opportunities it wouldn't get
by having a pseudo-random number generator do the work, but that
is not its point. Its point is metaphysical, not practical.

  In any case, *should* you want your faculty of practical reasoning to give different outputs for the very same inputs? Here we face

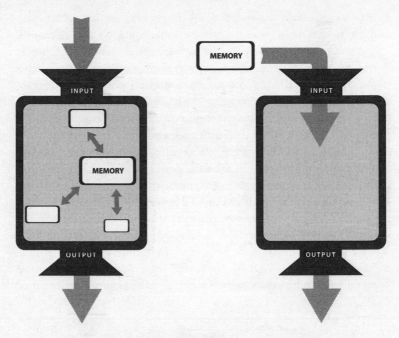

Figure 4.3 Memory Positions, Inside and Out

where is memory?

another boundary problem. What do we count as an input? Does the faculty contain the history of its previous activities, or is it just the content-free mill, the processor, which has to get (parts of) the history fed into it from external memory? (See Figure 4.3.)

You wouldn't want your practical reasoning to be so rigid that it made the same decision every day—for instance, always deciding on a ham sandwich for lunch. But if we include in the input available facts from memory, so that one of the inputs *today* is the fact that you've had a ham sandwich two days running, this makes today's case a different case from yesterday's case, however it is decided. Since people have capacious memories and perceptual sensitivity, they are never in exactly the same state twice, so they can get plenty of variability in the output of their faculties of practical reason by simply feeding in more varied input about their current state and circumstances. Your system of practical reasoning could be as reliable as a hand calculator, *determined* always to give output₁ in response to input₁ for every value of *i*, and yet still

→ det. to give certain output

[handwritten margin note top: → + still not make same decis]

[handwritten left margin: just chgs in input]

never make the same decision twice—simply because time marches on and the system never faces exactly the same input on two occasions. "That was then, this is now" as the saying goes. As we saw in Chapter 3, the computer chess programs playing against each other might never play the same game twice without ever adjusting *their* faculties of practical reason, all the variations being the result of changes in their inputs over time. You can be perfectly consistent and yet all over the map, if you let the features of the map influence your decision-making.

Now we are ready for Kane's central claim. Suppose your faculty of practical reasoning, unlike the deterministic arrangement just described, was equipped with indeterminism "somewhere between the input and the output." Is this a bug or a feature? How should we imagine this? Should we conceive of the faculty as containing one or more deterministic reasoning modules as subsystems, while also having some indeterministic innards? If we put a random number generator outside the faculty (Figure 4.4), then the random numbers it generates must be considered to be inputs to the faculty, and the faculty ought to treat

[handwritten left margin: indet. somewhere?]

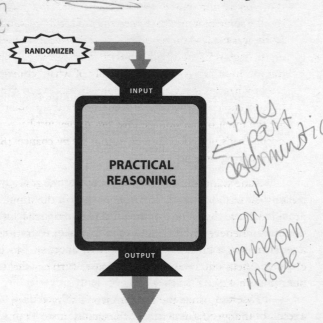

[handwritten right margin: this part ← deterministic ↓ or, random inside]

Figure 4.4   Randomizer Outside

them like any other inputs; if it is reliable, it should yield an output *determined* by that input. If, alternatively, we put a random number generator inside the faculty, to let it free up the way the faculty handles its inputs, then the faculty's outputs will *not* be determined by its inputs— but all we've done is drawn the boundary line in a different functional place.

Kane says that the indeterminacy should be "between" the input and the output, but we might well wonder why the indeterminacy couldn't come in *as part of* the input. What difference could it make? I put this question to Kane (in discussing an earlier draft of this chapter), and he had an interesting response:

> There is a reason why it is between input and output and does not come in merely as part of the input. The reason is that what is assumed to go on between input and output is the agent's doing or action (in the form of practical reasoning and efforts issuing in choice). The input (in the form of dispositions, beliefs, desires and the like) is not something the agent here and now controls, though some of it may have been the product of reasoning, efforts or choices made at earlier times. . . . Indeterminism merely at the input stage does not give us robust responsibility. The indeterminism must be an *ingredient* not only of what "comes to mind" but of what the agent is actually doing (reasoning, making efforts, making choices) to fully capture libertarian responsibility. If inputs are the result of our doings, OK, but if they just happen to us or occur, that's not good enough even if it's by chance. (Kane, personal correspondence)

Kane wants the indeterminism to be "the result of our doings" rather than randomness that "just happens" in the input. This is easily provided: Have the faculty of practical reasoning *send out for* some randomness whenever, in the midst of its labors, it encounters something it interprets as a blockade of one sort or another—an imponderable choice or meta-choice about which way to turn or what to think about next (Figure 4.5).

That way, since the randomness will have been "called for" as a result of the specific activities of the faculty, it won't just arrive unbidden from out of the blue. Moreover, the use to which the requested randomness gets put will be determined by constructive activities of

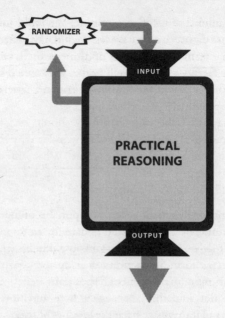

Figure 4.5 Sending Out for Randomness

the faculty itself. (If I decide to flip a coin to settle where to dine tonight, it is still my choice; I *made* it settle my choice.) But here again, we are just redrawing the boundary line; anything an *onboard* source of randomness can provide can also be provided in the *input* by an *external* source of randomness that is consulted when needed. As we are beginning to see, the metaphor of the container has to do a lot of work for Kane.

But, for the sake of argument, let's assume that Kane can come up with a good reason to distinguish internal from external sources of randomness. We install the indeterminacy inside the faculty, in between input and output, per his specifications, and then we install the faculty inside the agent. How does it operate in daily life? Kane notes that

> choices or decisions normally terminate processes of deliberation
> or practical reasoning, but they need not always do so. We need
> not rule out the possibility of impulsive, spur-of-the-moment, or
> snap, decisions, which also settle conditions of indecision but

impulsive
decis.
different

arise with minimal or no prior reasoning. Yet, while impulsive or snap decisions can occur, they are less important for free will than decisions that terminate processes of deliberation in which alternatives are reflectively considered. For, in the latter cases, we are more likely to feel we have control over the outcome and "could have done otherwise." (Kane 1996, p. 23)

So we get a picture of *occasional acts of* deliberate choice being the morally significant turning points—"they play a pivotal role" (p. 24)—laying down habits and intentions that are later acted on quite thoughtlessly but still with responsibility. Consider an example of a snap decision. My wife asks me if I can stop by the post office on my way to work and mail a package for her, and I reply almost instantaneously that I can't, because then I'd be late for an appointment with a student. Did I deliberate? Did I engage in a process of practical reasoning? This is not heavy-duty moral decision-making, but this is the stuff from which moral (and immoral) lives are largely composed: hundreds and thousands of minor choice points decided with a moment's consideration, usually with the background of justification kept tacit and unarticulated. How weird it would be if I had responded along these lines: "Well, since you are my wife and we have solemnly promised to help each other, and since I can think of no defect or problem in your request—you haven't asked me to do something physically impossible, or illegal, or self-destructive, for instance—there is undeniably a strong case for my answering, 'Yes, dear.' On the other hand, I have told a student that I would meet with him at nine-thirty, and given the traffic, honoring your request would entail standing him up for at least half an hour. I could try to call him and ask his permission to reschedule, but I might not reach him, and besides, the harder question is whether my mailing the package in so timely a manner is a sufficiently important errand to warrant inconveniencing him. My making the appointment amounted to a promise to him, though not one that couldn't be forgivably broken for cause. . . ." It is perhaps surprising to note that all these considerations (and many more!) really did contribute *somehow* to my snap answer. How so? Well, would I have given an unconsidered snap judgment, positive or negative, if my wife had asked me please to strangle the dentist on my way to work, or drive my car over a cliff? If I had earlier told my student merely that I intended to be in my office at 9:30 for coffee (no promise made or implied), or left the time of the appointment more flexible, or had been talking to him on the phone at the very moment my wife asked, this would have made a difference, surely, in my snap judgment. Even a snap judgment can be remarkably sensitive to myriad features of my world that have conspired over time to create my current dispositional state.

Kane is willing to allow that such a complex dispositional state, which has been building more or less continuously in me since I was a child, may *determine* how I will respond in such a case and in other cases when I do not deliberate. But once again, boundary questions loom. Should we view a snap judgment as *issuing from* the faculty of deliberation (but just so swiftly and effortlessly that the details stay tacit) or should we view the snap decision as issuing more directly from some "lower" faculty or subsystem, the faculty of deliberation being kept in reserve for occasional heavy lifting? It is best, I think, to draw the lines (which are, after all, just philosophers' lines of analysis, not anatomical boundaries to be discovered) so that even snap judgments get executed, effortlessly, in and by the faculty of practical reasoning. For, as we shall see, Kane holds that whereas the gap of indeterminism is to be located within that faculty (between input and output), the faculty does not always have to avail itself of indeterminism. It can operate deterministically on occasion, even when dealing with high-stakes moral decisions. (Shall I strangle the dentist? Naw.)

Kane is comfortable with this occasional role for determinism in the life of a moral agent, for several reasons. First, it permits him to handle these snap judgment cases realistically. It is just not plausible to maintain that the habits of a lifetime, yielding decisions so predictable you can trust your life to them, are nevertheless indeterministic (except in the limiting sense that there might be one chance in a bazillion that they would be disrupted). Think of your willingness to drive on the highway, facing oncoming cars in the opposite lane with an approach velocity well over 100 miles per hour. Your life depends on the drivers of those cars *not* deciding, as they are free to decide, to swerve suddenly into your lane, just to see what happens. Your equanimity on the highway shows how predictable you assume these total strangers to be. They *could* kill you in a senseless, suicidal *acte gratuit,* but you wouldn't pay a dollar or even a dime for the opportunity to clear the road of all oncoming cars before you ventured out. Second, Kane needs a helping of determinism in order to handle a more serious objection to libertarianism raised by me in *Elbow Room:* the case of Martin Luther.

> "Here I stand," Luther said. "I can do no other." Luther claimed that he could do no other, that his conscience made it *impossible* for him to recant. He might, of course, have been wrong, or have been deliberately overstating the truth. But even if he was—perhaps especially if he was—his declaration is testimony to the fact that we simply do not exempt someone from blame or praise for an act because we think he could do no other. Whatever Luther was doing, he was not trying to duck responsibility. (Dennett 1984, p. 133)

Kane accepts that Luther's decision was the furthest thing from a snap judgment, that it was definitely a morally responsible decision, and that what Luther said about it may well have been true: He could not have done otherwise; he truly was *determined by his faculty of practical reasoning at the time* to stand firm. The case of Luther is not a rare or unimportant sort of case. As we shall see in later chapters, the policy of preparing oneself for tough choices by arranging to be determined to do the right thing when the time comes is one of the hallmarks of mature responsibility, and Kane accepts this. In fact, he builds his account of free will around the idea that for each of us morally responsible agents, there must have been some relatively infrequent occasions in our lives when we have encountered conflicting desires—generating his type (iii) striving will. On some of these occasions we have decided to perform "self-forming actions" (SFAs), which may have a deterministic effect on our subsequent behavior, and only these SFAs need be the result of processes in the faculty of practical reason that are genuinely indeterministic:

> An act like Luther's can be ultimately responsible . . . though determined by his will, *because* the will from which it issued was a will *of his own making,* and in that sense it was his "own" free will. . . . Ultimately responsible acts, or acts done of one's own free will, make up a wider class of actions than those self-forming actions (SFAs) which must be undetermined and such that the agent could have done otherwise. But if no actions were "self-forming" in this way, we would not be *ultimately* responsible for anything we did. (Kane 1996, p. 78)

When I launch a boulder from a catapult toward my enemy, once the boulder is in flight, its trajectory is out of my hands, no longer subject to my will, but its effects on landing are my responsibility, no matter how long the delay. When I launch myself into a trajectory of one sort or another, having taken care to arrange that I will be unable to alter various aspects of that trajectory hereafter, the same conclusion manifestly holds. Reflections like this lead some libertarians to accept that the freedom they seek to install may have to be concentrated in a few windows of opportunity with special properties. (Peter van Inwagen, for instance, joins Kane on this point, but, unlike Kane, supposes such windows may be quite rare.) But now just what special properties will these be? Kane says that an SFA must meet condition AP:

> (AP) The agent has *alternative possibilities* (or can do otherwise) with respect to A at *t* in the sense that, at *t,* the agent *can* (has the *power* or *ability* to) do A and *can* (has the *power* or *ability* to) *do otherwise.* (Kane 1996, p. 33)

Notice the role of "at *t*" in this formula. Some philosophers can't bear to say simple things, like "Suppose a dog bites a man." They feel obliged instead to say, "Suppose a dog *d* bites a man *m* at time *t,*" thereby demonstrating their unshakable commitment to logical rigor, even though they don't go on to manipulate any formulae involving *d, m,* and *t.* Talk about time *t* is ubiquitous in philosophical definitions but seldom given any serious work to do. Here, however, it plays a serious role. This definition speaks about what is the case at each moment in time; it *requires* us to think about *possibilities-at-an-instant.* Kane (p. 87) quotes a rhapsodic passage from William James:

> The great point . . . is that the possibilities are really *here.* . . . At those soul-trying moments when fate's scales seem to quiver, . . . [we acknowledge] that the issue is decided nowhere else than *here* and *now. That* is what gives the palpitating reality to our moral life and makes it tingle . . . with so strange and elaborate an excitement. (James 1897, p. 183)

Let's look more closely at those quivering scales. Imagine that your faculty of practical reason is equipped with a dial, with a needle showing which way the scales are currently tipping as the mulling goes on, hovering between Go and Stay (supposing those are the options you're currently considering) and wandering back and forth, perhaps even quivering, oscillating swiftly between the two values (Figure 4.6). And suppose that at any moment you can terminate the process of deliberation by pressing the *Now!* button, sealing your choice with whatever side, Go or Stay, happens at that instant to be favored by the deliberation up to then. Suppose, for the moment, that all the processing by your faculty of practical reasoning is deterministic; it "sums the weights" by some deterministic function of all the input it has so far considered, and yields a moment-by-moment value that swings this way and that, between Go and Stay, depending on the order in which considerations are processed and reprocessed in the light of further deliberation.

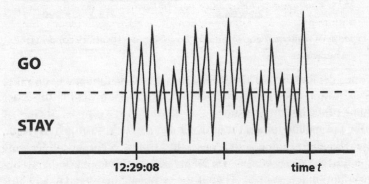

Figure 4.6   Quivering Needle between Go and Stay

Would condition AP be met in such a case? What would we look for to answer this question? Suppose we looked at the last *minute* of deliberation, and noticed that during that time, the needle oscillated back and forth a dozen times or more, and roughly half the time the needle pointed to Go and half the time the needle pointed to Stay. On that timescale it would certainly look as if both alternatives were *open* (compared, for instance, to a minute during which the needle rested firmly on Stay the entire time). But for Kane (and for James) this is not good enough. For there to be genuine free will, both possibilities have to open *at time t*, the very instant the *Now!* button was pressed. If we then zoomed in on that moment, and noticed that for the last 10 milliseconds before

time *t,* the needle was steady on Stay, which was also the decision registered by the pressing of the *Now!* button, it would seem that we had good
evidence that the Go option was *not* available at time *t* (see Figure 4.7).

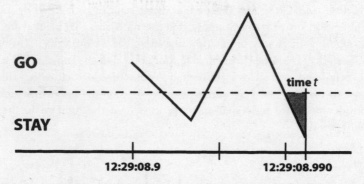

Figure 4.7   Enlargement of Figure 4.6, Showing 10-millisecond Period

Ah, but there is a loophole. I imagined that *you* got to press the
Now! button. Could we introduce indeterminacy by letting the exact
timing of the button pressing be "up to you"? Let's suppose, then, that
while the mulling process itself is all determined, what is indeterminate is the exact timing of the *Now!* button-press. Sometime in the next
20 milliseconds the button will be pressed, but exactly *when* is strictly
(quantum) indeterminate. Then if the quivering between Go and Stay
takes place at a high enough frequency to put both Go and Stay periods into that 20-millisecond window, the actual decision made by the
activation of the *Now!* button will be undetermined, utterly and offi-
cially unpredictable from a complete description of the universe at the
beginning of the window of opportunity (Figure 4.8).

Unfortunately, it will still not be the case that condition AP is
met, due to a flaw in the definition of AP: that pesky "at *t*" clause. It
will still be completely predictable that if the decision occurs at millisecond 5, say, it will be a decision to Go, and if it occurs at millisecond 17 it will be a decision to Stay. In fact, for any time *t* in the window
of opportunity, it is determined what decision would be made at that
instant; what isn't determined is when exactly the decision will be
made. The agent is not free *at t* to Go or Stay for *any* value of *t.* But
isn't this good enough, so long as the instant of choosing is undeter-

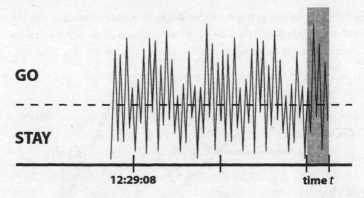

Figure 4.8   Window of Opportunity

[handwritten annotation: uncertainty of when decis rather than at time t are 2 optia]

mined? It is tempting to propose a mild revision of condition AP that would then accept our simple model: Let time *t* be smeared over the whole 20-millisecond time window instead of instantaneous, and we're home free, since both Go and Stay coexist at time *t* thus stretched out—and 20 milliseconds is hardly a long period of time.

The needle on the dial, and the button, make this model look awfully "mechanistic," to be sure, but Kane demands this himself. He's trying to be a *naturalist* libertarian, so he wants his model to be scientifically respectable, something the brain could implement, and the dial and the button are just vivid devices for helping us visualize the underlying state of the relevant neural complexity. *Some* sort of physically realizable neural state must implement the current weighting, and *some* state-transition must implement a decision (yield an output); we can just pretend that the dial transduces the former and the button triggers the latter. So the model illustrates one way—one family of ways—in which subatomic quantum indeterminacy could be amplified into playing a crucial role in decision-making. Moreover, the model seems to satisfy Kane's Ultimacy requirement for SFAs:

(U) for every X and Y (where X and Y represent occurrences of events and/or states) if the agent is personally responsible for X, and if Y is an *arche*[5] (or sufficient ground or cause or explanation)

5. *Arche* is Aristotle's term for *origin*.

for X, then the agent must also be personally responsible for Y.
(Kane 1996, p. 35)

Translation: You can only be personally responsible for one thing if you
are personally responsible for everything that is a sufficient condition
for it. According to Kane,

> SFAs are the undetermined, regress–stopping voluntary actions
> (or refrainings) in the life histories of agents that are required if U
> is to be satisfied. (Kane 1996, p. 75)

Indeterministic timing of the *Now!* button could make the decision
itself indeterministic in cases where both options quiver in a slightly
elongated window of opportunity; there wouldn't *be* any sufficient
condition for either the Go or Stay decision at any earlier moment, so
you could be personally responsible for Go (or Stay) without having
to worry about being responsible for any earlier sufficient condition
for Go (or Stay). Of course, we still have to find some way of making
sense of an indeterministic button pressing being "up to *you*" and not
itself just an *external,* random input.

"If you make yourself really small, you can externalize virtually everything"[6]

Once again we have a boundary problem, and this time it is major: How
can Kane get a quantum indeterminacy to be *inside* the relevant system?
To see the difficulty, suppose a bystander yells just as you're about to
push the *Now!* button, startling you and thereby hastening your press
by five milliseconds, *causing* your press. Is the decision now no longer
yours at all? After all, the crucial part of the cause, the part that determined whether to Go or Stay, was itself caused by the bystander's yell
(which was caused by the seagull flying by so close, which was caused
by the early return of the fishing fleet, which was caused by the resump-

---

6. This was probably the most important sentence in *Elbow Room* (Dennett 1984, p. 143),
and I made the stupid mistake of putting it in parentheses. I've been correcting that mistake
in my work ever since, drawing out the many implications of abandoning the idea of a
punctate self. Of course, what I meant to stress with my ironic formulation was the converse:
You'd be surprised how much you can internalize, if you make yourself large.

break causation

tion of El Niño, which . . . was caused by a butterfly flapping its wings back in 1926). Even if that butterfly wing flap was truly undetermined, the magnified effect of a quantum leap in its tiny brain, this moment of indeterminism is *in the wrong time and place.* The butterfly's moment of freedom back in 1926 isn't what gives you free will today, is it? Kane's libertarianism requires him to break the chain of causation somewhere *in* the agent and *at* the time of decision, the "here and now" requirement spoken of so eloquently by William James. If it really matters, as libertarians think, then we'd better shield your processes of deliberation from all such *external* interference. We'd better insulate the wall that surrounds . . . *you* so that external forces don't interfere with the decision you're cooking up in your internal kitchen, using only the ingredients that *you* have allowed through the door.

This retreat of the Self into a walled enclave within which all the serious work of authorship has to be done parallels another retreat into the center of the brain, the various misbegotten lines of argument and reflection that lead to what I call the Cartesian Theater, the imaginary place in the center of the brain "where it all comes together" for consciousness. *There is no such place,* and any theory that tacitly presupposes that there is should be set aside at once as on the wrong track. All the work done by the imaginary homunculus in the Cartesian Theater must be distributed *in time and space* in the brain. The problem is compounded for Kane, since he has to figure out some way to get the undetermined quantum event to be not just *in you* but *yours.* He wants above all for the decision to be "up to you," but if the decision is undetermined—the defining requirement of libertarianism—it isn't determined by you, whatever you are, because it isn't determined by anything. Whatever you are, you can't *influence* the undetermined event—the whole point of quantum indeterminacy is that such quantum events are not influenced by anything—so you will somehow have to *co-opt* it or *join forces* with it, putting it to use in some intimate way, an *objet trouvé* that you meaningfully incorporate into *your* decision-making in some fashion. But in order to do this, there has to be more to you than just some mathematical point; you have to *be someone;* you have to have parts—memories, plans, beliefs, and desires—that you've acquired along the way. And then all those causal influences from the past, from outside, come crowding back in, contaminating the workshop, preempting your creativity, usurping control of your decision-making. A serious quandary.

The problem, you will recall, was already clearly recognized by William James when he asked, "If a 'free' act be a sheer novelty, that comes not *from* me, the previous me, but *ex nihilo,* and simply tacks itself on to me, how can *I,* the previous I, be responsible?" Kane makes some useful headway on an answer to this rhetorical question with his idea of "plural rationality" (Kane 1996, Chapter 7). We don't want our free acts to be unmotivated, inexplicable, random lightning bolts without rhyme or reason. We want there to be reasons for them, we want these to be our reasons, *and* (if we're libertarians) we want them to meet the AP condition, to be free in the sense that "at time *t*" we "could have done otherwise." One way this could be the case is if you yourself have taken the time and effort to develop two (or more) sets of *competing* reasons. Then both sets of reasons are composed, devised, revised, sanded, and polished *locally,* by you yourself. Though you may have borrowed some pieces and ideas from outside, you've made them your own, so these are indeed *do-it-yourself* reasons. Moreover, each set of reasons is at least tentatively endorsed by you. (If one of them wasn't, there wouldn't have been any fuss, would there? You'd have made a quick—perhaps even snap—decision in favor of the other.) So when deliberation finally terminates, whichever side you come down on is a side you have taken very seriously yourself, right up to the verge of endorsement. Your act amounts to a final verdict, a declaration that makes you the kind of person you are (a Stayer or a Goer)—and *right then* you could have done otherwise.

The point of plural rationality—or "parallel processing," as he more recently calls it (Kane 1999)—is that it builds on an intuition we've always had: You can be rightly held responsible for the outcome of a deed that includes a chance or undetermined element, *if that is what you were trying to accomplish.* The would-be assassin whose lucky long shot hits the prime minister is not absolved on the grounds that it was mere chance—even genuinely indeterministic chance—that he hit his target. By setting up an opponent process pitting two different attempts against each other (e.g., the businesswoman's quandary about whether to do the right thing or advance her career), Kane guarantees that when one of the attempts fails, the other succeeds, and she is rightly held responsible in either case because *that is one of the things she was trying to accomplish.* The fact that she was trying to accomplish two incompatible things at the same time doesn't show that when she manages to accomplish one of them, she *wasn't* trying to accomplish it!

So Kane claims that this embedding of indeterminism in the maelstrom of conflicting reasons, where the agent is actually *trying*—type (iii), the striving will—to get it right, saves the outcome, whichever it is, from being a fluke, a mere accident. Every adult agent will have faced such dilemmas, moral or prudential, and been shaped by them.

> By choosing one way or another in such cases, the agents would
> be strengthening their moral or prudential characters or reinforc-
> ing selfish or imprudent instincts, as the case may be. They would
> be "making" themselves or "forming" their wills one way or
> another in a manner that was not determined by past character,
> motives, and circumstances. . . . It is because their efforts are thus
> a response to inner conflicts embedded in the agents' prior char-
> acter and motives that their character and motives can explain the
> conflicts and why the efforts are being made, without also explain-
> ing the outcomes of the conflicts and the efforts. Prior motives
> and character provide reasons for going either way, but not deci-
> sive reasons explaining which way the agent will inevitably go.
> (Kane 1996, p. 127)

The idea that someone who has been tested by serious dilem-mas of practical reasoning, who has wrestled with temptations and quandaries, is more likely to be "his own man" or "her own woman," a more responsible moral agent than someone who has just floated happily along down life's river taking things as they come, is an attractive and familiar point, but one that has largely eluded philosophers' attention. In most accounts of free will, the occurrence of tough choices in an agent's history plays no marked role and, in fact, is largely ignored, probably because it draws attention to the embarrassing limiting case: Buridan's Ass, who purportedly starves to death because he is equidistant from two piles of food and can't think of a reason for going left rather than right (or vice versa). This "liberty of indifference" has been noted since medieval times, and tie-breaking by flipping a coin has always been a recognized solution to such impasses, a useful prosthesis of the will, one might say, but it doesn't look like a good model for free will. If we theorists find ourselves approaching a view in which

[handwritten: Character building → Free will]

our only free choices will be those where we might as well flip a coin, then we must have blundered down the wrong path. Turn back quickly. And so the topic gets ignored. But Kane shows quite convincingly that the incremental character-building that *may* (but also may not) grow out of a lifetime of hard choices taken seriously really does add a "variety of free will worth wanting." There's one big problem with it, however: It doesn't need the indeterminism that inspired its creation. Moreover, it can't harness indeterminism in any way that distinguishes it from determinism, because the "here and now" requirement is not only not well motivated; it is also probably incoherent, as we shall see.

[handwritten: but doesn't need indet.]

Beware of Prime Mammals

The basic idea is that the ultimate responsibility *lies where the* ultimate cause *is.*

—Robert Kane, *The Significance of Free Will*

You may think you're a mammal, and that dogs and cows and whales are mammals, but really there aren't any mammals at all—there couldn't be! Here's a philosophical argument to prove it (drawn, with alterations, from Sanford 1975).

(1) Every mammal has a mammal for a mother.

(2) If there have been any mammals at all, there have been only a finite number of mammals.

(3) But if there has been even one mammal, then by (1), there have been an infinity of mammals, which contradicts (2), so there can't have been any mammals. It's a contradiction in terms.

Since we know perfectly well that there are mammals, we take this argument seriously only as a challenge to discover what fallacy is lurking within it. Something has to give. And we know, in a general way, what has to give: If you go back far enough in the family tree of any mammal, you will eventually get to the therapsids, those strange, extinct bridge species between the reptiles and the mammals. A gradual transition occurred from clear reptiles to clear mammals, with a lot of hard-to-classify intermediaries filling in the gaps. What should we

[handwritten: don't draw lines of gradual chg]

do about drawing the lines across this spectrum of gradual change? Can we identify a mammal, the Prime Mammal, that didn't have a mammal for a mother, thus negating premise (1)? On what grounds? Whatever the grounds are, they will be indistinguishable from the grounds we could also use to support the verdict that that animal was *not* a mammal—after all, its mother was a therapsid. What should we do? We should quell our desire to draw lines. We don't need to draw lines. We can live with the quite unshocking and unmysterious fact that, you see, there were all these gradual changes that accumulated over many millions of years and eventually produced undeniable mammals.

Philosophers tend to like the idea of stopping a threatened infinite regress by identifying something that is—must be—*the* regress-stopper: the Prime Mammal, in this case. It often lands them in doctrines that wallow in mystery, or at least puzzlement, and, of course, it commits them to essentialism in most instances. (The Prime Mammal must be whichever mammal in the set of mammals first had all the *essential* mammalian features. If there is no definable essence of *mammal,* we're in trouble. And evolutionary biology shows us that there are no such essences.)

Kane's theory of free will specifically calls for "regress-stopping" special cases, the self-forming acts, or SFAs.

> If an infinite regress is to be avoided, there must be actions somewhere in the agent's life history for which the agent's predominant motives and the will on which the agent acts were *not already set one way.* (Kane 1996, p. 114)

[handwritten: → not already set]

One might pause to ask how often these important moments tend to occur. Once a day on average, or once a year or once a decade? Do they tend to start at birth, at age five, at puberty? These SFAs look suspiciously like Prime Mammals. It is worrying that while they are key events in the life of any moral agent—the natural rites of passage, one might say, into responsible adulthood—they are practically impossible to discover. There is no way to tell a genuine SFA from a pseudo-SFA, an impostor bout of reasoning that never *actually* availed itself of quantum indeterminism but just cranked out a pseudo-random and hence deterministic result. They would feel the same from the inside and look the same from the outside, no matter how sophisticated our observational apparatus. As Paul Oppenheim has suggested to me, Kane's SFAs can be usefully compared with *speciation events* in evolution, which can only be retrospectively identified. Every birth in every lineage is a potential speciation event, since offspring all have at least minute differences that make them unique, and any difference could be the beginning of something that eventually blooms into speciation. Time will tell. There is nothing *special* at the time about a birth that will turn out to have been a speciation event.[7] Similarly, one should be suspicious of the demand that there be an event—an SFA—that has some special, intrinsic, local feature that sets it apart from its nearest kin and explains its capacity to found something important. Is it plausible that an agent who hadn't yet experienced one or more of these very special events (but only near misses, pseudo-SFAs) would simply not be responsible for any acts performed? "Yes, these furry, warm-blooded things *look* a lot like mammals, and smell and sound like mammals, and are cross-fertile with mammals, but they lack the secret essence; they aren't mammals at all, not really."

Consider Luther in this regard. Kane says: "If he is ultimately accountable for his present act, then at least some of these earlier choices or actions must have been such that he could have done otherwise with respect to them. If this were not the case, nothing he could have ever done would have made any difference to what he was" (Kane 1996, p. 40). And so it makes sense—one might think—to take a good hard look at Luther's biography, to see what kind of upbringing he had, what powerful influences held him in thrall, what catastrophes he endured, and the like. But, in fact, nothing we could discover about such macroscopic details would shed *any light at all* on the question of whether or not Luther had had any genuine SFAs during this period. We could certainly discover that episodes of conflict and soul-searching occurred on various occasions, and we might even confirm that these

---

7. Some contemporary creationists have conceded that all living things are related by descent in a tree of life that is billions of years old, and also grant that all the transformations of successive generations within species are accomplished by mindless Darwinian natural selection, but hold out hope that the branching events themselves, the speciations, are, if not miraculous, in need of special help from some intelligent designer (or Intelligent Designer—they claim to be neutral about the identity of the i.d.). This condensation of all the specialness into a magic moment—or a place where it all comes together—is an irresistible motif to some thinkers. The clearest example is Michael Behe (1996); for a discussion of the fallacies involved, see Dennett (1997C).

occasions set up "chaotic" opponent processes in the neural networks from which his decisions eventually emerged. What we could not discover, however, was whether these tugs-of-war had the benefit of genuinely random, as opposed to mere pseudo-random, sources of variability. The price libertarians must pay for sequestering their pivotal moments in subatomic transactions in some privileged place in the brain (at time t) is that they render these all-important pivots undetectable by both the everyday biographer and the fully equipped cognitive neuroscientist. One might think that the difference between Luther$_1$, who was held in a cell during his adolescence for five years and subjected to brainwashing, and Luther$_2$, who had a roughly normal adolescence of triumphs and trials in the knockabout world, would have a bearing on whether there were SFAs in the ancestry of the decision made by Luther$_{today}$. But these salient environmental differences, which intuitively *do* have a bearing on our assessment of Luther's capacity for moral choice, are in no way symptoms of the presence or absence of SFAs. (They are just as irrelevant to the question of whether or not an SFA occurred in Luther as Austin's ten demonstration putts would be to the question of whether or not he was determined to miss the putt at time t.) And when we get out our supermicroscopes and look at subatomic activity in the neurons, whatever we see will be equally uninformative about SFAs.

But isn't this inscrutability of ultimate responsibility a problem for every theory? As Kane has said,

> If a young murderer is on trial and we look into his past life of child abuse and peer pressure, we have to make some judgment about how much of his present vicious character from which this act flowed is his own doing and how much is due to outside influences over which he lacked control. Such questions are relevant to determining guilt or innocence and how much punishment should be mitigated on any theory. They are formidably difficult questions to answer no matter what view you take about free will. (Kane, personal correspondence)

This is right, so far as it goes. Variations in life history are indeed relevant to variations in current degree of responsibility, as Kane says, and they are also difficult to investigate, on any theory. But Kane's libertarian view requires an additional investigation that is hard to

motivate—impossible, in my opinion. Consider the situation statistically: We sort a hundred murderers by background, from most deprived
to most fortunate, to see which should have mitigation, or total exculpation (we'll address those policy issues later). Suppose we find the following: 60 percent show *clear* evidence of major deprivation of the
relevant sorts and are hence unproblematic candidates for substantial
mitigation; 10 percent are "borderline"—they show quite a lot of deprivation, but how much is too much?—and the remaining 30 percent
show normal-to-exemplary upbringings, no signs at all of brain damage, etc. (See Figure 4.9.) These fortunate individuals emerge, by a
process of elimination, as practically indistinguishable from each other
in all the macroscopic characters that we take to be the necessary conditions for responsibility—the features the 60 percent lack. They are
all *apparently* responsible adults. They are all among society's apparent
success stories—we raised them right, filled in their gaps, gave them
an equal opportunity, and so forth.

Nature doesn't insist on sharp boundaries, but sometimes *we*
must draw a line of political policy, simply because we have to have
some practical and ostensibly fair way of dealing with specific cases: You

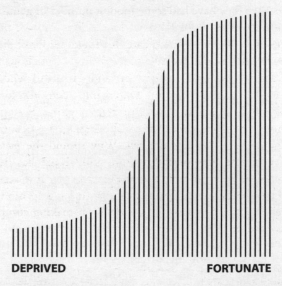

DEPRIVED   **FORTUNATE**

Figure 4.9   The Distribution of Murderers

can't drive until you are sixteen in most states, and you can't drink until you are twenty-one, no matter how mature you are for your age. Faced with the array of cases illustrated in Figure 4.9, we would have to find some partly arbitrary way of drawing a line across the penumbral 10 percent, and opinions would no doubt differ on which factors to weight heavily and which to ignore. (If the curve were much steeper, we'd be grateful to discern an apparent joint at which to carve nature; if it were more gradual, our political task would be all the harder.) But Kane's view requires us to reserve judgment about not just the 10 with marginal claims to mitigation but the 30 exemplary candidates as well. Some unknown number—it could be all 30—could turn out to be *entirely* nonresponsible, because all the apparent SFAs in their life histories were pseudo-SFAs. After all, Kane holds that no robot with only a pseudo-random number generator in its system could be responsible *at all,* and yet such a robot might pass all macroscopic tests for humanity perfectly. (Such a robot, unlike a Stepford Wife,[8] would not betray its robotitude by a slavish obsession with some one policy, thanks to the pseudo-random jigglers in its faculty of practical reason that would keep it eternally open-minded.) Indeed, according to Kane's view it is entirely possible that some in the marginal group of 10 are rightly held responsible since they have had some modest number of genuine SFAs in their pasts, in spite of their deprivations, while some of the privileged group of 30 are not fit candidates for moral responsibility at all.

Try to imagine the first defendant (the son of a billionaire, since he'll need an expensive team of lawyers and scientists!) who tries to introduce evidence in court before sentencing, "demonstrating" that his brain lacked the quantum indeterminacies required for responsibility, even though he'd had an exemplary upbringing, was of above average intelligence, etc. It's a tough sell. Why should the *metaphysical* feature of Ultimate Responsibility (supposing Kane has defined a coherent possibility) count more than the macroscopic features that can be defined independently of the issue of quantum indeterminism, and that are well motivated in terms of the decision-making competences that agents have or lack? Indeed, why should metaphysical Ultimate

8. The 1975 science fiction movie *The Stepford Wives,* by Bryan Forbes (based on Ira Levin's novel), portrayed a town in which the real wives were gradually replaced by mindless robot duplicates who devoted all their energy to housecleaning and taking care of their men.

[handwritten annotations: "ultimate Resp- → not a variety of free will worth wanting" ... "looks like chance"]

Responsibility count for anything at all? If it can't be motivated as a grounds for treating people differently, why should anyone think it is a variety of free will worth wanting? As Kane himself puts it, "In short, when described from a physical perspective alone, *free will looks like chance*" (Kane 1996, p. 147). And chance looks exactly the same, whether it is genuinely indeterministic or merely pseudo-random or chaotic.

The libertarian, like the essentialist in biology, is captivated with boundaries, in particular the boundaries that delimit the "here" and the "now." But these boundaries, being partly interdefinable, are porous in any case. Suppose the indeterministic neurons in your faculty of practical reasoning died, leaving you disabled for any future SFAs. But suppose, fortunately for you, that the damaged part of your brain could be replaced by an indeterministic prosthetic device implanted in just the right milieu in the healthy part of your brain. A good way to get genuine quantum indeterminism into a physical device is to use a little bit of decaying radium and a Geiger counter, but it might not be healthy to have such a radium randomizer implanted in your brain, so it could be left in the lab, surrounded by a lead shield, and its results could be fed into your brain on demand, by radio link (as in my "Where am I?" story in *Brainstorms,* 1978). The location of the randomizer in the lab obviously shouldn't make a difference, since it is *functionally* inside the system; it would play exactly the same role as the damaged neurons used to play, no matter where it was geographically. But there might be a cheaper, safer way of getting exactly the same effect: We could use genuinely random fluctuations in the light coming from deep space as our trigger, beaming it directly to the transceiver implanted in your brain. Since this signal arrives at the speed of light, there is no way for us to predict what the next fluctuations will be, even though their random source is a star light-years away. But if there is no problem getting your indeterminacy from a distant star, why insist on making it *now* in the first place? *Record* a series of random fluctuations by a radium randomizer over a century, and install that recording from the past as your pseudo-random number generator somewhere in your brain, to be consulted when appropriate.

In *Elbow Room,* I noted the unimportance of the difference between a lottery in which the winning ticket is chosen (randomly) *after* all the tickets are sold, and a lottery in which the winning ticket stub is chosen *before* the tickets are sold. Both are fair lotteries; both give all the purchasers a fair chance of winning.

> If our world is determined, then we have pseudo-random number generators in us, not Geiger counter randomizers. That is to say, if our world is determined, all our lottery tickets were drawn at once, eons ago, put in an envelope for us, and doled out as we needed them through life. (Dennett 1984, p. 121)

Kane has suggested to me (personal correspondence) that "The indeterminacy-producing mechanism must be responsive to the dynamics within the agent's own will and not override them or it would be making the decisions and not the agent." His concern is that a remote source of randomness would threaten your autonomy, and be likely to take control of your thinking processes. Wouldn't it be much safer—and hence more responsible—to keep the randomizer inside you, under your watchful eye in some sense? No. Randomness is just randomness; it isn't *creeping* randomness. Programmers routinely insert calls to the random number generator in their programs, not worrying about it somehow getting out of hand and providing chaos where it isn't wanted. Suppose we visualize the brain's dynamics in our Go/Stay example as creating a saddle in a decision landscape, a place where the decision-explorer will eventually slide off the hill into either the Go valley to the north or the Stay valley to the south. (See Figure 4.10.)

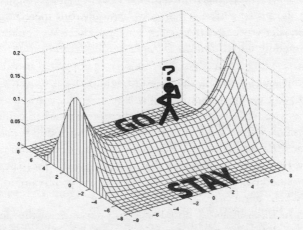

Figure 4.10 Saddle in a Decision Landscape

The landscape is generously sprinkled with banana peels—calls to the random number generator that are activated any time the decision-explorer passes over them. This keeps the explorer moving, randomly if necessary, preventing Buridan's Ass from occurring, so the explorer never gets stuck on the flattish ridge of the saddle and dies decisionless. These slippery banana peels are harmless, though, because once a decision starts heading down into one valley or the other, encountering an unnecessary peel can only briefly bump the decision back uphill a bit, delaying for a micro-moment the plunge that has already been settled on, or else hasten its downward slide, without being able to overrule it. Or to use another vivid image popular among modelers, the random number generator simply "shakes" or "jiggles" the landscape ever so incessantly, so that nothing can just stop on the saddle forever—but the shape of the landscape isn't altered at all, so nothing ominous "takes over."

How Can It Be "Up to Me"?

A popular argument with many variations claims to demonstrate the incompatibility of determinism and (morally important) free will as follows:

(1) If determinism is true, whether I Go or Stay is completely fixed by the laws of nature and events in the distant past.

(2) It is not up to me what the laws of nature are, or what happened in the distant past.

(3) Therefore, whether I Go or Stay is completely fixed by circumstances that are not up to me.

(4) If an action of mine is not up to me, it is not free (in the morally important sense).

(5) Therefore, my action of Going or Staying is not free.

Kane's libertarian response to this compelling argument is to attempt to isolate the indeterminism of libertarian free will in a few crucial episodes of possibility "at time t" and he hopes to locate those episodes inside the agent, both spatially and temporally, so the agent's choices

han far back do you go? (handwritten note)

can be "up to" the agent. But once he has allowed that the morally relevant effects of these episodes can be widely distributed in time (as in
the case of Luther), what work is there left for the boundary of the container to do? If some event in Luther's boyhood can play a crucial role
in Luther's responsibility in adulthood for his momentous decision not
to recant, why not an event in Luther's mother's life while Martin was
but an embryo? Because, presumably, those events occurred not in
Luther but outside Luther, in the external environment, however
strongly they imposed themselves on him, and hence they were not "up
to Luther." Yes, but if "the child is father to the man," isn't young
Luther just as external to adult Luther? Why aren't Luther's youthful
dispositions, and even his later conscious episodic memories of his
youth, themselves rather remote influences "from the outside"? This is
a stretched version of the problem we encountered early in this chapter, when we wondered whether to put the memory inside the faculty
of practical reasoning or leave it outside and have portions of it
"inputted" when the occasion demanded it. The lines we draw don't
do any discernible work for us. And as we will see later, our own moral
agency often depends crucially on a little help from our friends without in any way being thereby diminished. The ideal of "do-it-
yourself," carried to absolutistic extremes, is superstition. It is true that
if you make yourself as small as possible, you can externalize virtually
everything. So much the worse for models that push all that matters
into a single moment, somewhere in the heart of an atom. If there is
a case to be made for libertarianism, it will have to come from some
still unexplored quarter, since the best attempt to date, Kane's, ends up
in a cul-de-sac. His Ultimate Responsibility requirement turns out, on
further examination, to burden the *specs* of a free agent with conditions
that are both unmotivated and undetectable. You can demand a car
with two steering wheels and a compass in the gas tank, but that doesn't
make it worth wanting.

   How then should we respond to the incompatibilist argument? Where is the misstep that excuses us from accepting the conclusion? We can now recognize that it commits the same error as the
fallacious argument about the impossibility of mammals. Events in the
distant past were indeed not "up to me," but my choice now to Go
or Stay is up to me because its "parents"—some events in the *recent*
past, such as the choices I have recently made—were up to me

chain of "up to mes"

(because *their* "parents" were up to me), and so on, not to infinity, but far enough back to give my *self* enough spread in space and time so that there is a *me* for my decisions to be up to! The reality of a moral me is no more put in doubt by the incompatibilist argument than is the reality of mammals.

Before leaving the topic of libertarianism, we should ask, once more, what the point of it might be. An indeterministic spark occurring at the moment we make our most important decisions couldn't make us more flexible, give us more opportunities, make us more self-made or autonomous in any way that could be discerned *from inside or outside,* so why should it matter to us? How could it be a difference that makes a difference? Well, it could be, could it not, that belief in such a spark, like belief in God, changes the whole way you think about the world and your life in it, even if you'll never know (in this lifetime) whether it is true. Yes, the case for belief in indeterminism in action must come down to something like that. But there is an important difference. Even if you can never know, never prove scientifically, that there is a God, it is not hard to explain why a belief in a supreme and merciful Being watching over you might comfort you, give you moral strength and hope, and so forth. The belief in God is not like, say, the belief in Gog (a large sphere of copper that orbits a star outside our light-cone and has the letters GOG stamped prominently on its surface). Anybody is welcome to believe in Gog if it makes them feel good, but why would it? My charge is that libertarians have inflated perfectly reasonable desires for varieties of free will worth wanting into a craving for a variety of free will that would be no more worth wanting than communion with Gog. But it is also true that however misguided such a craving is, it might be unwise to tamper with it. It might be that until or unless a suitable substitute is found, we should tiptoe away from further criticism of this irrational and unmotivated yearning. *(Stop that crow!)* But if that is so, it's too late to put the cat back in the bag. We'd better see what can be done to help people get over their delusion.

Chapter 4

An examination of the best positive case for libertarianism shows that it cannot find a defensible location for indeterminism within the decision-making processes

→ no dispensible loc. of indetermination

of a responsible agent. Since it cannot motivate its defining requirement, we can leave indeterminism behind and consider more realistic requirements for freedom, and how they could have evolved.

More realistic req. for freedom

Chapter 5

Four billion years ago, there was no freedom on our planet, because there was no life. What kinds of freedom have evolved since the origin of life, and how did evolutionary reasons—Mother Nature's reasons—evolve into our reasons?

Notes on Sources and Further Reading

I drew the importance of chaos to philosophers' attention in *Elbow Room* (Dennett 1984). A more recent compatibilist appreciation of the role of chaos is Matt Ridley 1999, pp. 311–13. On where the buck stops, see *Elbow Room* (p. 76), which also includes discussions of Newtonian chaos (pp. 151–52) and the movable clutch that marks the difference between weakness of will and self-deception.

The discussion of snap judgments in the faculty of practical reason is a descendant of the discussion of *getting a joke* that I offered in *Brainchildren* (Dennett 1998A, p. 86): The complex dispositional state of belief that determines whether or not one will laugh at a joke depends on one's filling in many details left unsaid in the telling. It would be odd to call the unconscious process that triggers an involuntary chuckle *deliberation,* but it is a sophisticated information-transforming process in any case.

See David Velleman's "What Happens When Someone Acts?" (1992) on Chisholm's agent causation, and a possible reduction of it to something more acceptable to a naturalist, a topic taken up in Chapter 8 of this book.

Theorists seldom explicitly endorse the Cartesian Theater, but closet Cartesians can sometimes be teased into the open. For a collection of examples on display, with commentary, see my recent books and articles on consciousness. A similar image of isolation for the sake of authorship inspires, and distorts, some philosophers' thinking about *understanding.* See my "Do-it-yourself Understanding," in *Brainchildren* (Dennett 1998A), on Fred Dretske's attempt to save genuine home-made understanding from pre-fab simulacra that can be bought and installed on the cheap. (According to this vision, robots may *seem* to understand, but it isn't *their* understanding, since they didn't make it themselves.)

On Kane's idea of parallel processing: In a piece entitled "On Giving Libertarians What They Say They Want" (in Dennett 1978) I made much the same suggestion, using the example (pp. 294–95) of a woman who had to choose between taking a job at the University of Chicago, and taking a job at Swarthmore; either decision is rational, and even if the choice is undetermined, when she makes whichever choice she makes, there is a good reason for it, and it is *her* reason. But I didn't take the idea very seriously, except as a crumb to throw to libertarians. Kane shows that I underestimated it.

On mammals: There is quite a literature that has grown up in recent years on vagueness and how to deal with it. I recommend in particular Diana Raffman (1996); she has convinced me, but if her discussion doesn't convince you, you can follow her bibliographical references to the rest.

Robert French's (1995) Tabletalk model is a deeply satisfying architecture for the sort of stochastic decision-making process sketched here—a toy world without moral significance, but full of insight. See my Foreword to his book, reprinted in *Brainchildren* (Dennett 1998A).

Kane proposes a distinction between what he calls "Epicurean" and "non-Epicurean" versions of indeterminism (Kane 1996, pp. 172–74). A world of Epicurean indeterminism consists of "forks in history" (modeled on the Epicureans' random swerves) interspersed among things and events with "determinate" properties. In a non-Epicurean world, there is "both indeterminateness of physical properties and the possibility of forks in history." What difference does this make? "An Epicurean world in which undetermined events occurred given an entirely determinate past—a world of chance without indeterminacy—would be a world of mere chance, not free will. There would be no indeterminate 'gestation period' for free acts, so to speak; they would just pop out of a determinate past one way or the other without any preparation in the form of indeterminacy-producing tension, struggle, and conflict" (p. 173). But what about the computer models of non-linear, chaotic, recurrent feedback tugs-of-war? They have *apparent* "gestation periods" as pregnant with (digital approximations of) indeterminacy as you like, but they get their (pseudo-)indeterminism the Epicurean way—with pseudo-random number generators interspersing their outputs among the deterministic subroutines. You can't have it both ways: If, following Paul Churchland, you want to applaud the discovery of the power of non-linear, recurrent networks, in all their non-symbolic, non-rigid, free-wheeling holistic openness, you have to concede that Epicurean algorithmicity suffices to provide it, since that is what the working models are made of.

Chapter 5

WHERE DOES ALL THE DESIGN COME FROM?

"Excuse me, sir, can you tell me how to get to Symphony Hall?"
"Practice, practice, practice!"

The Boston Symphony Orchestra is notorious for giving guest conductors a hard time until they prove themselves. A young conductor, facing his debut with the BSO and knowing their reputation, decided to try a shortcut to respect. He was scheduled to conduct the premiere of an unhearably discordant contemporary piece, and as he reviewed the score a brilliant stratagem occurred to him. He found an early crescendo in which the entire orchestra was screaming away on more than a dozen different quarreling notes and noted that the second oboe, one of the softest voices in the orchestra, was scheduled to play a B-natural. He picked up the part score for the second oboe, and carefully inserted the sign for a flat—the oboe would now be instructed to play B-flat. At the first rehearsal, he briskly led the orchestra up through that doctored crescendo. "No!" he hollered, stopping the orchestra abruptly. Then, with furrowed brow and deep concentration, he said, "Somebody, let's see, yes, it must be . . . second oboe. You were supposed to play B-natural and you played B-flat." "Hell, no," said the second oboe, "I played B-natural. Some idiot had written in a B-flat!"

Early Days

Consider this phenomenon from the biological point of view. The Boston Symphony Orchestra has been in existence for more than a century, its personnel continuously being replaced, its finances waxing and waning, its repertory growing and shifting as old chestnuts are retired and new pieces explored. In many ways this fine old institution is like a living organism, with a distinct personality, a particular history of growth, of sickness and health, learning and forgetting, traveling around the globe and returning to its home, replacing tired old "cells" with new recruits, adjusting its behavior to the ecological niche in which it flourishes. This biological perspective is compelling, and useful, but it leaves out the most amazing and important features of the phenomenon. If biologists from another galaxy were to discover the Boston Symphony Orchestra, what ought to impress them most are not these remarkable similarities to animals and plants, but the dissimilarities. An organism is made of a huge team of cells, but no cell can be anxious about the prospect of being humiliated. No cell can learn to play the oboe, or be responsible for choosing this year's guest conductors from a list of young hopefuls. No cell can draw out the implications of the oboist's response and anticipate the catastrophic effect it will have on the young conductor's campaign for respect.

What is remarkable about the Boston Symphony Orchestra (and the myriad other human institutions and practices) is that, on the one hand, they can be so beautifully designed and organized, so self-sustaining, while, on the other hand, they are composed of a motley assortment of *autonomous* individuals, of different nationalities, ages, genders, temperaments, aspirations. The orchestra members are free to come and go as they choose, so the board of directors must work hard to ensure that the working conditions and pay are sufficient to keep the orchestra members well motivated. Look at the violin section. Twenty talented individuals, but all different. Some are brilliant but lazy while others are obsessive perfectionists; one is bored but conscientious, another is enraptured by the music, yet another is daydreaming about making love to that adorable cellist over there, but all of them are drawing their bows across their strings in perfect unison, a pattern robustly superimposed on a kaleidoscope of different human consciousnesses. What makes this concerted action possible is a massive complex of cultural products, deeply shared by the musicians, the audience, the composer, the conservatories, the banks, the municipal authorities, the violin-makers, the ticket agencies, and so on. Nothing in the animal world is a close counterpart to this complexity. Human minds are furnished—and beset—by thousands of anticipations, evaluations, projects, schemes, hopes, fears, and memories that are entirely inaccessible to the minds of even our closest relatives, the great apes. This world of human ideals and artifacts gives individual human beings capacities and proclivities that are strikingly different from those of any other living beings on the planet.

The freedom of the bird to fly wherever it wants is definitely a kind of freedom, a distinct improvement on the freedom of the jellyfish to float wherever it floats, but a poor cousin of our human freedom. Compare birdsong to human language. Both are magnificent products of natural selection, and neither is miraculous, but human language revolutionizes life, opening up the biological world in dimensions utterly inaccessible to birds. Human freedom, in part a product of the revolution begat of language and culture, is about as different from bird freedom as language is different from birdsong. But in order to understand the richer phenomenon, one must first understand its more modest components and predecessors. What we must do to understand human freedom is to follow Darwin's "strange inversion of reasoning" and go back to a time at the beginning of life when there was no freedom, no intelligence, no choice, but only proto-freedom, proto choice, proto intelligence. We have already reviewed in outline what happened: Simple cells eventually begat complex cells, which eventually begat multicellular organisms, which then begat the complex macroscopic world we live and act in. Now we must go back and look at some of the telling details in this procession.

Suppose you just want to be alive on planet Earth. What do you need? Starting at the molecular level, you need not just DNA, but all the molecular tools—proteins—for accomplishing the many steps in DNA replication. You need one protein for initiating the process, another for unwinding the helix, another for binding the single-strand DNA, . . . , relaxation of supercoils, chromosome segmentation/packing, and so on. None of them is optional; all of them are necessary. If you're missing any of these proteins you're out of luck. These building blocks themselves had to be designed over time. The complete kit, which we share with all life on the planet today, got assembled and refined over several billion years, and it replaces simpler kits for our still simpler ancestors. We are dependent on our kit, and they were dependent on theirs, but we have more possibilities than they did, because the

↳ proteins came from DNA

improvements in our kit made possible higher forms of aggregation, and these in turn made possible ever more devious ways of colliding with the other things in the world, and exploiting the results of those collisions. When life began, there was just one way of being alive. It was do A or die. Now there are options: do A or B or C or D or . . . die.

To live you need energy. Did the first energy exploited for life come from the sun, or from thermal sources deep in the earth? This is currently an open question, with a tantalizing array of hypotheses about the origins of life competing for confirmation. However it got started, life—most of it, in any case—eventually became dependent on energy from the sun. To stay alive and reproduce you had to float on or near the surface of the sea, basking in sunlight. A major innovation occurred when some of the baskers mutated, "discovering" thereby that instead of doing it all themselves, they could do better by engulfing and disassembling some of their neighbors, using them as a handy store of fancy spare parts already constructed. Encroachment is what makes life interesting. Encroachers and encroachees inaugurated an arms race, leading to new varieties of both. Soon—in a billion years or so—there were many "ways of making a living" (as Richard Dawkins has put it), but these many ways will always be but a Vanishing thread of actuality in the Vast space of logical possibility. Almost every combination of building blocks is a way of not being alive.

One of the most important innovations in this arms race of competitive design was the accident known as the eukaryotic revolution, which happened some billion years ago. The first living things, the relatively simple cells known as prokaryotes, had the planet to themselves for around three billion years, until one of them got invaded by a neighbor, and the resulting team-of-two was more fit than their uninfected cousins, so they prospered and multiplied, passing their teamwork on to their offspring. It was an early instance of a sort of cooperation: *symbiosis,* a case in which X and Y collide, but instead of X destroying Y, or vice versa, or even worse, mutual self-destruction— the usual result of collisions in this hard world—X and Y join forces, creating Z, a new, bigger, more versatile thing, with better options. This may have happened many times in the prokaryotic world, of course, but once it happened the first time, the planet was changed for all subsequent life. These super-cells, eukaryotes, lived alongside their prokaryotic cousins, but were enormously more complex, versatile, and competent thanks to their hitchhikers. This was unwitting cooperation, of course. The eukaryotic teams were utterly oblivious of the teamwork in which they engaged. They had—and needed—no appreciation of the free-floating rationale for their advantage over the competition. The early eukaryotes were not themselves multicellular, but they opened up the design space of multicellular organisms since they had enough spare parts to become different kinds of specialists. (We're still a long way from violinists and oboists, and the teamwork of the BSO, but we're on the road.)

The eukaryotic revolution draws our attention to the fact that even in biological evolution, which Darwin aptly called "descent with modification," there is plenty of room for *horizontal* transmission of design. The prokaryote hosts who were first "infected" by their symbiotic visitors got a huge gift of competence *designed elsewhere*. That is, they didn't get all their competence by *vertical* descent from their ancestors via their parents and grandparents and so forth. They didn't get all their competence from their genes, in other words. They did, however, pass on this gift to all their offspring and grand-offspring through their genes, since the genes of the invaders came to share the fate of the nuclear genes of their hosts, traveling side by side into the next generation, which was infected at birth, one might say, with its own complement of symbionts. The clear trace of this dual path is still highly salient today, in all multicellular creatures, including us. Mitochondria, the tiny organelles that transform energy in each of our cells, are the descendants of such symbiont invaders, and have their own genomes, their own DNA. Your mitochondrial DNA, which you get only from your mother, exists in each of your cells, alongside your nuclear DNA—your genome. (Sexual reproduction came along later; the sperm from your father contributed none of his mitochondria in the process of fertilization.)

Horizontal transmission of design, of information that can be put to good uses, is the key feature of human culture, and undoubtedly the secret of our success as a species. Each of us is the beneficiary of the design work done by countless others who are not our ancestors. We don't each have to "reinvent the wheel" or invent calculus or clocks or the sonnet form. It is sometimes claimed, erroneously, that this cultural transmission, being between genetically unrelated individuals, shows that human culture cannot be interpreted as an evolutionary phenomenon governed by the principles of neo-Darwinian theory. In fact, as we have just seen, horizontal transmission of good design elements between unrelated individuals is recognized as an important feature of evolution of early (single-celled) life, with a growing list of proven instances, a centerpiece, not an embarrassment, of contemporary evolutionary biology.

The eukaryotic revolution was not accomplished overnight; solutions to many problems had to be laboriously discovered by evolution before it was secure. In Chapter 2, we met the parasitic transposons, renegade genes whose deleterious effects had to be thwarted. The process that resolved these intragenomic conflicts illustrates several important Darwinian themes: R&D is expensive, every design must be "paid for," and evolution is forever reusing earlier designs (paid for and copied) for new purposes. Simple prokaryotes can get their genes *expressed* with relatively simple gene-reading equipment. That is, it doesn't take very high tech to follow a prokaryote gene recipe and build an offspring prokaryote. Fancier eukaryotic cells, however, to say nothing of us multicellular types composed of these more complex building blocks, need a mind-bogglingly elaborate system of intermediate steps, checks and balances, so that genes can be turned on and off at appropriate times by the indirect effects of other gene products, and so forth. For some time biologists have had a classic chicken-and-egg puzzle to contend with: How did this elaborate gene-regulation machinery evolve? Multicellular life couldn't even begin to evolve until most of this expensive machinery was in place, but it apparently isn't required for simpler prokaryotic life. What paid for all that R&D? The answer that is now emerging is that it was paid for by a civil war that raged for roughly a billion years of early prokaryotic life. It was an arms race within the genome, with good-citizen genes doing battle with those transposons—freeloaders who copied themselves repeatedly in the genome without providing any benefit to the whole organism. This created lots of measures and countermeasures, such as silencing mechanisms and isolation-defeating mechanisms. (The details of these mechanisms, like the details of the mechanisms that permitted the symbiotic unifications of genomes in the eukaryotic revolution, are beginning to emerge, and are fascinating, but well beyond the scope of this book.) Like modern-day arms races, the result was an expensive standoff, but the fruits of that R&D were then available for beating into plowshares: the high-tech machinery necessary for making multicellular life forms (McDonald 1998). So it appears that we ourselves are a "peace dividend" of sorts, like computers and Teflon and GPS, and the other high-tech spin-offs of the arms race conducted by the military-industrial complex thanks to our tax dollars.

The Prisoner's Dilemma[1]

But how do these arms races actually work? What factors govern or constrain the thrust and counterthrust of the different "sides" in these competitions? Every circumstance in nature in which something like *cooperation* arises requires explanation. (It may well begin with a happy accident, but it can't be sustained as a happy accident. That would be too good to be true.) This is where we need the perspective of game theory, and its classic example, the Prisoner's Dilemma. This is a simple two-person "game" that casts shadows, both obvious and surprising, into many different circumstances in our world. Here is the basic scenario. You and another person have been imprisoned pending trial (on a trumped-up charge, let's say), and the prosecutor offers each of you, separately, the same deal: If you both hang tough, neither confessing nor implicating the other, you will each get a short sentence (the state's evidence is not that strong); if you confess and implicate the other and he hangs tough, you go scot-free, and he gets life in prison; if you both confess and implicate, you both get medium-length sentences. Of course, if you hang tough and the other person confesses, he goes free and you get life. What should you do?

     If you both could hang tough, defying the prosecutor, this would be much better for the two of you than if you both confess, so couldn't you just promise each other to hang tough? (In the standard jargon of the Prisoner's Dilemma, the hang-tough option is called *cooperating*—with the other prisoner, of course, not the prosecutor.) You could promise, but you would each then feel the temptation—whether or not you acted on it—to *defect,* since then you would go scot-

1. Parts of this section are drawn, with revisions, from *Darwin's Dangerous Idea* (Dennett 1995, pp. 253–54).

free, leaving the *sucker,* sad to say, in deep trouble. Since the game is symmetrical, the other person will be just as tempted, of course, to make a sucker of you by defecting. Can you risk life in prison that the other will keep his promise? Probably safer to defect, isn't it? That way, you definitely avoid the worst outcome of all, and might even go free. Of course, the other fellow will figure this out, too, if it's such a bright idea, so he'll probably play it safe and defect, too, in which case you *must* defect to avoid calamity—unless you are so saintly that you don't mind spending your life in prison to save a promise-breaker—so it is likely that you'll both wind up defecting and accepting medium-length sentences. If only you both could overcome this reasoning and cooperate!

The logical structure of the game is what matters, not this particular setting, which is a usefully vivid imagination-driver. We can replace the prison sentences with positive outcomes (it's a chance to win different amounts of cash or, say, descendants) just so long as the payoffs are symmetrical, and ordered so that lone defection pays more than mutual cooperation, which pays each more than mutual defection does, which in turn pays more than the sucker payoff one gets when the other is a lone defector. (And in formal settings we set a further condition: The average of the Sucker and Mutual Defection payoffs must not be greater than the Mutual Cooperation payoff.) *Whenever this structure is instantiated in the world, there is a Prisoner's Dilemma.*

Game-theoretic explorations have been undertaken in many fields, from philosophy and psychology to economics and biology. In evolutionary game theory, the payoffs are measured in descendants, and the point of the models is to explore the conditions under which

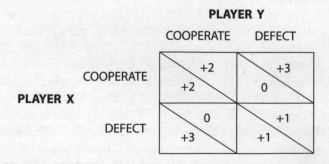

Figure 5.1   The Prisoner's Dilemma

"cooperative" designs can hold their own and outproduce the otherwise always favored selfish defectors. Why is defection the default winning strategy? Look at the payoff matrix in Figure 5.1. Whatever Player Y does, if Player X defects, he will do better than if he cooperates. Defection is said to *dominate* as a policy in the basic situation. The effect of this on Player X's descendants as a proportion of the next generation of a population can be derived mathematically, and readily demonstrated in simulations, in which simple defector-agents of one sort or another are matched pairwise with simple cooperator agents of one sort or another. They interact according to their type—defectors always defect and cooperators always cooperate—and the outcomes (in terms of numbers of descendants) are tallied and summed over many generations. In the absence of special features to prevent it, the defectors soon swamp the cooperators, sad to say. This ineluctable trend is the prevailing wind against which all evolution of cooperation must be pitched. The most influential of the many applications of game-theoretic thinking to evolutionary theory is John Maynard Smith's concept of an *evolutionarily stable strategy, or ESS,* a strategy that may not be the best imaginable but is un-subvertible by any alternative strategy under the circumstances. The nasty world in which everybody defects all the time is an ESS in most imaginable circumstances, since pioneer cooperators thrown into such a population get suckered to death in short order. There are conditions, however, in which there are other, more heartening outcomes, and these escapes from the grim default are the steps in the ladder leading up to us.

There can be no doubt that game-theoretic analyses work in evolutionary theory. Why, for instance, are the trees in the forest so tall? For *the very same reason* that huge arrays of garish signs compete for our attention along commercial strips in every region of the country! Each tree is looking out for itself and trying to get as much sunlight as possible. If only those redwoods could get together and agree on some sensible zoning restrictions and stop competing with each other for sunlight, they could avoid the trouble of building those ridiculous and expensive trunks, stay low and thrifty shrubs, and get just as much sunlight as before! But they can't get together. Under these circumstances, defection from any cooperative agreement is bound to pay off whenever it occurs, so if there weren't an essentially inexhaustible supply of sunshine, trees would be stuck with the "tragedy of the commons"

(Hardin 1968). The tragedy of the commons occurs when there is a finite "public" or shared resource that individuals will be selfishly tempted to take more of than their fair share—such as the edible fish in the oceans. Unless very specific and enforceable agreements can be reached, the result will tend to be the destruction of the resource. It was the evolution of enforceable checks and balances that permitted the cooperating genes to hold their own against the defecting transposons, one of the earliest "technological" innovations to overcome the boringly simple world of universal selfishness, universal defection.

E Pluribus Unum?[2]

The advent of multicellularity was ushered in by another innovation in cooperation: solving the problem of group solidarity at the cell level. As I noted at the beginning of Chapter 1, we are each of us composed of trillions of robotic cells, each with its own complete set of genes and an impressive array of internal life-support machinery. Why do these individual cells submit so selflessly to the good of the whole team? They have *become* hugely dependent on each other, of course, and cannot survive long on their own except in the particular environment they usually inhabit, but how did they get that way?[3] One of the virtues of the "gene's-eye perspective" on evolution is that it draws attention to this issue as a serious problem. Cellular group solidarity is ubiquitous in nature; slavishly devoted cells can be found, after all, in every living thing visible to the naked eye. Hence it is "natural," but it is

2. This section contains revised versions of a section with the same title in Chapter 16 of *Darwin's Dangerous Idea* (Dennett 1995).
3. Notice that I have fallen in with the standard biologists' way of speaking about biological *types* (or lineages or species) as if I were talking about individuals. Our cells have "become dependent" but none of *my* cells have *become* dependent; they were all born that way. Giraffes have grown longer necks over eons, and it took thousands of years for weaverbirds to "learn" how to build their nests. The "growing" and "learning" here is invisible if you concentrate on individuals. As we saw with the emergence of avoidance in Chapter 2, even when each individual is determined to be just the way it is till the day it dies, the larger process can yield change, improvement, growth. Some philosophers have been suspicious of this duality of perspective—"bait and switch" is how I characterized their skepticism in *Darwin's Dangerous Idea* (Dennett 1995)—but it is the key to understanding how all evolutionary R&D happens.

nonetheless a design achievement of major proportions, not something that biologists may take for granted. The lessons to be learned are tricky, however, because the cells that compose us belong to two very different categories.

The cells that compose multicellular me all share an ancestry; they are a single lineage, the daughter cells and granddaughter cells of the egg and sperm that united to form my zygote. They are *host* cells. The other cells, the *symbionts,* are the same sort of things—they are themselves eukaryotes and prokaryotes—but they count as outsiders because they have descended from different lineages. (So this is second-generation symbiosis; symbiosis created your eukaryotic cells, which have then played host in turn to a flood of newer guests!)

What difference does the host/guest difference make? The answer here, which will be echoed at the higher level of human social life, is that although pedigree is often a good predictor of future competence, it is future competence that counts in the end, regardless of pedigree. For instance, your immune system is composed of cells that are now members in good standing of the host team, but they began their career in your ancestors as an invading army, which was gradually co-opted and turned into a troop of mercenary guards, their own genetic identity merged with that of the more ancient lineages they joined forces with, another instance of horizontal transmission of design. The key to understanding the patterns these transformations follow is to treat all these robotic cells as tiny individual agents, as intentional systems, each with a smidgen of "rational" decision-power. Adopting the intentional stance, leaping up from the physical stance of component atoms, via the design stance of simple machines, to the intentional stance of simple agenthood, is a tactic that pays off handsomely but must be used with caution. It is all too easy to miss the fact that there are crucial moments in the careers of these various agents and semi-agents and hemi-semi-demi-agents when opportunities to "decide" arise, and then pass.

The cells that compose my bulk have a shared fate, but some in a stronger sense than others. The DNA in my finger cells and blood cells is in a genetic cul-de-sac; these cells are part of the *somatic* line (the body), not the *germ* line (the sex cells). As François Jacob has memorably said, the dream of every cell is to become two cells, but my somatic line cells are doomed to die "childless"—aside from occasionally yielding replacements for neighbors who die in action, and barring dramatic progress in cloning techniques. Since this cul-de-sac was determined some time ago, there is no longer any pressure, any normal opportunity, any "choice points" at which their intentional trajectories—or the trajectories of their limited progeny—might be adjusted. They are, you might say, *ballistic* intentional systems, whose highest goals and purposes have been fixed once and for all, with no chance of reconsideration or guidance. They are totally committed slaves to the *summum bonum* of the body of which they form a part. They may be exploited or tricked by visitors, but under normal circumstances they cannot rebel on their own. Like the Stepford Wives, they have a single *summum bonum* designed right into them, and it is not "Look out for Number One." On the contrary, they are team players by their very nature.

How they further this *summum bonum* is also designed right into them, and in this regard they differ fundamentally from the other cells that are "in the same boat": my symbiont visitors. The benign mutualists, the neutral commensals, and the deleterious parasites that share the vehicle they all together compose—namely, me—each have their own *summum bonum* designed into them, and it is to further their *own* respective lineages, not mine. Fortunately, there are conditions under which an *entente cordiale* can be maintained, for after all, they are all in the same boat, and the conditions under which they can do better by not cooperating are limited. But they do have the "choice." It is an issue for them in a way it is not, normally, for the host cells.

Why? What enables—or requires—the host cells to be so committed, while giving the visitor cells a free rein to rebel when the opportunity arises? Neither sort of cell is a thinking, perceiving, rational agent, of course. And neither sort is significantly more cognitive than the other. That is not where the fulcrum of evolutionary game theory is located. Redwood trees are not notably clever either, but they are in conditions of competition that force them to defect, creating what is, from *their* point of view (!) a wasteful tragedy. The mutual cooperative agreement whereby they would all forgo growing tall trunks in the vain attempt to gain more than their fair share of sunlight is evolutionarily unenforceable.

The condition that creates a choice is the mindless "voting" of *differential* reproduction. It is the opportunity for differential reproduction that lets the lineages of our visitors "change their minds" or "reconsider" the choices they have made, by "exploring" alternative policies. My host cells, however, have been designed once and for all by a single vote at the time my zygote was formed. If, thanks to mutation, dominating or selfish strategies occur to *them,* they will not flourish (relative to their contemporaries), since there is scant opportunity for differential reproduction. (Cancer can be seen as a selfish—and vehicle-destructive—rebellion made possible by a revision in normal circumstances that does permit differential reproduction.)

Brian Skyrms has pointed out (1994A, 1994B) a wonderful parallel between this multicellular policy (another benign fruit of the civil war that created all the gene-reading machinery) and John Rawls's monumental *A Theory of Justice* (1971). The precondition for normal cooperation in the strongly shared fate of somatic-line cells is analogous to the situation in "the original position," Rawls's thought experiment about how rational agents would choose to design an ideally just state, if they had to choose from behind what he called the veil of ignorance. Skyrms calls this, aptly, the "Darwinian Veil of Ignorance." Your sex cells (sperm or ova) are formed by a process unlike that of normal cell division, or *mitosis.* Your sex cells are formed by a different process called *meiosis,* a process that randomly constructs a *half* a genome-candidate (to join forces with a half from your mate) by choosing first a bit from "column A" (the genes you got from your mother) and then a bit from "column B" (the genes you got from your father) until a full complement of genes—but just one copy of each—is constructed and installed in a sex cell, ready to try its fate in the great mating lottery. But which daughters of your original zygote are destined for meiosis and which for mitosis? This too is a lottery.

Is it a *random* lottery or a *pseudo-random* lottery? So far as we know it is just like a coin toss, *determined* by some inscrutable and unpatterned coincidence of impingements from who knows where, and hence predictable in principle by Laplace's *infinite* demon, but not by the highly sensitive and broadly based selective forces that form the blind but effective gropings of the Blind Watchmaker. Thanks to this mechanism, paternal and maternal genes (in you) could not ordinarily "know their fate" in advance. The question of whether they are going to have germ-line progeny that might have a flood of descendants flowing on into the future or be relegated to the sterile backwaters of somatic-line slavery for the good of the body politic or corporation (think of the etymology) is unknown and unknowable, so there is nothing to be gained by selfish competition among their fellow genes. That, at any rate, is the usual arrangement. There are special occasions, however, on which the Darwinian Veil of Ignorance is briefly lifted: the cases of "meiotic drive" or "genomic imprinting" (Haig and Grafen 1991; Haig 1992, 2002; for discussion, see *Darwin's Dangerous Idea* [Dennett 1995, Chapter 9]), in which circumstances *do* permit a "selfish" competition between genes to arise—and arise it does, leading to escalating arms races. But under most circumstances, the "time to be selfish," for genes, is strictly limited, and once the die—or the ballot—is cast, those genes are just along for the ride until the next election. The parallel was perhaps first noted by E. G. Leigh:

> It is as if we had to do with a parliament of genes: each acts in its own self-interest, but if its acts hurt the others, they will combine together to suppress it. The transmission rules of meiosis evolve as increasingly inviolable rules of fair play, a constitution designed to protect the parliament against the harmful acts of one or a few. However, at loci so closely linked to a distorter that the benefits of "riding its coattails" outweigh the damage of its disease, selection tends to enhance the distortion effect. Thus a species must have many chromosomes if, when a distorter arises, selection at most loci is to favor its suppression. Just as too small a parliament may be perverted by the cabals of a few, a species with only one, tightly linked chromosome is an easy prey to distorters. (Leigh 1971, p. 249)

Just try to describe these deep patterns in nature without using the intentional stance! The slow-motion patterns that are predictive at the gene level are remarkably reminiscent of—actually previews of—the patterns that are predictive at the psychological and social level: opportunities, discernment and ignorance, seeking out the best moves against the competition, avoidance and retaliation, choice and risk. The moves and countermoves in evolutionary R&D have rationales even if nothing and no one explicitly considered them. These are what I call *free-floating rationales,* and they preceded our articulated, considered rationales by billions of years. Among them is the fundamental principle of avoidance of harm, the same in both domains: When you don't

[handwritten margin note: don't have inf of fate, no avail of free choice]

have any information about what your fate is likely to be, you cannot avail yourself of a free choice.

> And this is another way of denying people an opportunity: keeping them in the dark about it. We might call such an unrecognized and unimagined opportunity a *bare opportunity*. If I walk by a row of trash cans, and one of them happens to contain a purse full of diamonds, then I pass up a bare opportunity to become wealthy. . . . Bare opportunities are in great abundance, but they are not enough; when we say we want opportunities, or chances to improve our lots, we don't want just bare opportunities. We want to detect our opportunities, or be informed about them, in time to act. (Dennett 1984, pp. 116–17)

[handwritten margin note: → coordin.]

Skyrms shows that when the individual elements of a group—whether of whole organisms or their parts—are closely related (clones or near clones) or are otherwise able to engage in mutual recognition and assortative "mating," the simple Prisoner's Dilemma, in which the strategy of defection always dominates, does not correctly model the circumstances. That is why our somatic cells don't defect; they are clones. This is *one* of the conditions under which groups—such as the group of my "host" cells—can have the harmony and coordination required to behave, quite stably, as an "organism" or "individual." But before we give three cheers and take this to be our model for how to make a just society, we should pause to notice that there is another way of looking at these model citizens, the somatic-line cells and organs: Their particular brand of selflessness is the unquestioning obedience of zealots, exhibiting a fiercely xenophobic group loyalty that is hardly an ideal for human emulation.

We, unlike the cells that compose us, are not on ballistic trajectories; we are *guided* missiles, capable of altering course at any point, abandoning goals, switching allegiances, forming cabals and then betraying them, and so forth. For us, it is always decision time. For this reason, we are constantly faced with social opportunities and dilemmas of the sort for which game theory provides the playing field and the rules of engagement, but not yet the solutions. Life is more complicated for people in society than it is for the cells that compose them, and we have a lot of R&D to accomplish—"Practice, practice, practice!"—before we get to Symphony Hall.

trial + error?

Still, it is heartening to recognize that the problems facing *us* have precedents that were eventually solved by trial and error. Otherwise we wouldn't be here. Trial and error—even mindless trial and error, with preservation of partial progress—is a potent process. It has created genuine novelties in the world; it has solved major problems, overcome daunting obstacles. Trial and error works, so *trial* works: At least one variety of trying has a proven track record. Our varieties of trying may not look quite so feckless in the face of determinism when we see how successful their ancestors have been. The very cells that compose us are the direct descendants of cells that once had to solve a huge problem of cooperation, and succeeded.

Digression: The Threat of Genetic Determinism

With all this ominous talk about cells and genes in juxtaposition with talk about violinists and oboists, it is time, perhaps, to set minds at ease by raising the "specter" of "genetic determinism" and banishing it once and for all. According to Stephen Jay Gould, genetic determinists believe the following:

> If we are programmed to be what we are, then these traits are ineluctable. We may, at best, channel them, but we cannot change them either by will, education, or culture. (Gould 1978, p. 238)

If this is genetic determinism, then we can all breathe a sigh of relief: There are no genetic determinists. I have never encountered anybody who claims that will, education, and culture cannot change many, if not all, of our genetically inherited traits. My genetic tendency to myopia is canceled by the eyeglasses I wear (but I do have to *want* to wear them); and many of those who would otherwise suffer from one genetic disease or another can have the symptoms postponed indefinitely by being *educated* about the importance of a particular diet, or by the *culture-borne* gift of one prescription medicine or another. If you have the gene for the disease phenylketonuria, all you have to do to *avoid* its undesirable effects is stop eating food containing phenylalanine. As we have seen, what is inevitable doesn't depend on whether or not determinism reigns, but on whether or not there are steps we can take, based on information we can get in time to take those steps, to avoid the fore-

doesn't dep on determinism

seen harm. There are two requirements for a meaningful choice: information and a path for the information to guide. Without one, the other is useless or worse. In his excellent survey of contemporary genetics, Matt Ridley (1999) drives the point home with the poignant example of Huntington's disease, which is "pure fatalism, undiluted by environmental variability. Good living, good medicine, healthy food, loving families or great riches can do nothing about it" (p. 56). This is in sharp contrast to all the equally undesirable genetic predispositions that we *can* do something about. And it is for just this reason that many people who are likely, given their family tree, to have the Huntington's mutation choose *not* to take the simple test that would tell them with virtual certainty whether they have it. But note that if and when a path opens up, as it may in future, for treating those who have the Huntington's mutation, these same people will be first in line to take the test.

Gould and others have declared their firm opposition to "genetic determinism," but I doubt if anybody thinks our genetic endowments are infinitely revisable. It is all but impossible that I will ever give birth, thanks to my Y chromosome. I cannot change this by either will, education, or culture—at least not in my lifetime (but who knows what another century of science will make possible?). So at least for the foreseeable future, some of my genes fix some parts of my destiny without any real prospect of exemption. If that is genetic determinism, we are all genetic determinists, Gould included. Once the caricatures are set aside, what remains, at best, are honest differences of opinion about just how much intervention it would take to counteract one genetic tendency or another and, more important, whether such intervention would be justified. These are important moral and political issues, but they often become next to impossible to discuss in a calm and reasonable way. A first step toward restoring sanity is to recognize, as a useful rule of thumb, that whenever you encounter the "charge" of "genetic determinist" the likelihood is high that this is just a case of *Stop that crow!* and doesn't warrant any more discussion, at least not in those terms. Besides, what would be so specially bad about *genetic* determinism? Wouldn't *environmental* determinism be just as dreadful? Consider a parallel definition of environmental determinism:

> If we have been raised and educated in a particular cultural environment, then the traits imposed on us by that environment are

[handwritten marginalia: determinism → inevitability]

ineluctable. We may, at best, channel them, but we cannot change them either by will, further education, or by adopting a different culture.

The Jesuits have often been quoted (I don't know how accurately) as saying: "Give me a child until he is seven, and I will show you the man." An exaggeration for effect, surely, but there is little doubt that early education and other major events of childhood can have a profound effect on later life. There are studies, for instance, that suggest that such dire events as being rejected by your mother in the first year of life increases your likelihood of committing a violent crime (e.g., Raine et al. 1994). Again, we mustn't make the mistake of equating determinism with inevitability. What we need to examine empirically—and this can vary just as dramatically in environmental settings as in genetic settings—is whether the undesirable effects, however profound, however large, can be avoided by steps we can take. Consider the affliction known as *not knowing a word of Chinese*. I suffer from it, thanks entirely to environmental influences early in my childhood (my genes had nothing—nothing directly—to do with it). If I were to move to China, however, I could soon enough be "cured," with some effort on my part, though I would no doubt bear deep and unalterable signs of my deprivation, readily detectable by any native Chinese speaker, for the rest of my life. But I could certainly get good enough in Chinese to be held responsible for actions I might take under the influence of Chinese speakers I encountered.

[handwritten marginalia: Chapter Luck is that just environment?]

Isn't it true that whatever isn't determined by our genes must be determined by our environment? What else is there? There's Nature and there's Nurture. Is there also some X, some further contributor to what we are? There's Chance. Luck. We've seen, in Chapters 3 and 4, that this extra ingredient is important but doesn't have to come from the quantum bowels of our atoms or from some distant star. It is all around us in the causeless coin-flipping of our noisy world, automatically filling in all the gaps of specification left unfixed by our genes, and unfixed by salient causes in our environment. This is particularly evident in the way the trillions of connections between cells in our brains are formed. It has been recognized for years that the human genome, large as it is, is much too small to *specify* (in its gene recipes) all the connections that are formed between neurons. What happens is that the genes specify processes that set in motion huge population growths of neurons—many times more neurons than our brains will eventually use—and these neurons send out exploratory branches, at random (at pseudo-random, of course), and many of these *happen* to connect to other neurons in ways that are *detectably* useful (detectable by the mindless processes of brain-pruning). These winning connections tend to survive, while the losing connections die, to be dismantled so that their parts can be recycled in the next generation of hopeful neuron growths a few days later. This selective environment within the brain (especially within the brain of the fetus, long before it encounters the outside environment) no more *specifies* the final connections than the genes do; saliencies in both genes and developmental environment influence and prune the growth, but there is plenty that is left to chance.

When the human genome was recently published, and it was announced that we have "only" about 30,000 genes (by today's assumptions about how to identify and count genes), not the 100,000 genes that some experts had surmised, there was an amusing sigh of relief in the press. Whew! "We" are not just the products of our genes; "we" get to contribute all the specification that those 70,000 genes would otherwise have "fixed" in us! And how, one might ask, are *we* to do this? Aren't we under just as much of a threat from the dread environment, nasty old Nurture with its insidious indoctrination techniques? When Nature and Nurture have done their work, will there be anything left over to be *me*? (If you make yourself really small, you can externalize virtually everything.)

Does it matter what the trade-off is if, one way or another, our genes and our environment (including chance) divide up the spoils and "fix" our characters? Perhaps it seems that the environment is a more benign source of determination since, after all, "we can change the environment." That is true, but we can't change a person's *past* environment any more than we can change her parents, and environmental adjustments in the future can be just as vigorously addressed to undoing prior genetic constraints as prior environmental constraints. And we are now on the verge of being able to adjust the genetic future almost as readily as the environmental future. Suppose you know that any child of yours will have a problem that can be alleviated by either an adjustment to its genes or an adjustment to its environment. There can be many valid reasons for favoring one treatment policy over another, but it is certainly not obvious that one of these options should be ruled out on moral or metaphysical grounds. Suppose, to make up an imaginary case that will probably soon be outrun by reality, you are a committed Inuit who believes life above the Arctic Circle is the only life worth living, and suppose you are told that your children will be genetically ill-equipped for living in such an environment. You can move to the tropics, where they will be fine—at the cost of giving up their *environmental* heritage—or you can adjust their genomes, permitting them to continue living in the Arctic world, at the cost (if it is one) of the loss of some aspect of their "natural" genetic heritage.

The issue is not about determinism, either genetic or environmental or both together; the issue is about *what we can change* whether or not our world is deterministic. A fascinating perspective on the misguided issue of genetic determinism is provided by Jared Diamond in his magnificent book *Guns, Germs, and Steel* (1997). The question Diamond poses, and largely answers, is why it is that "Western" people (Europeans or Eurasians) have conquered, colonized, and otherwise dominated "Third World" people instead of vice versa. Why didn't the human populations of the Americas or Africa, for instance, create worldwide empires by invading, killing, and enslaving Europeans? Is the answer . . . *genetic?* Is science showing us that the ultimate source of Western dominance is in our genes? On first encountering this question, many people—even highly sophisticated scientists— jump to the conclusion that Diamond, by merely addressing this question, must be entertaining some awful racist hypothesis about European genetic superiority. So rattled are they by this suspicion that they have a hard time taking in the fact (which he must labor mightily to drive home) that he is saying just about the opposite: The secret explanation lies not in our genes, not in *human* genes, but it does lie to a very large extent in genes—the genes of the plants and animals that were the wild ancestors of all the domesticated species of human agriculture.

Prison wardens have a rule of thumb: If it can happen, it will happen. What they mean is that any gap in security, any ineffective prohibition or surveillance or weakness in the barriers, will soon enough be found and exploited to the full by the prisoners. Why? The intentional stance makes it clear: The prisoners are intentional systems who are smart, resourceful, and frustrated; as such they amount to a huge

domest. of plants/ animls

supply of informed desire with lots of free time in which to explore their worlds. Their search procedure will be as good as exhaustive, and they will be able to tell the best moves from the second-best. Count on them to find whatever is there to be found. Diamond exploits the same rule of thumb, assuming that people anywhere in the world have always been just about as smart, as thrifty, as opportunistic, as disciplined, as foresighted, as people anywhere else, and then showing that indeed people have always found what was there to be found. To a good first approximation, *all the domesticable wild species have been domesticated.* The reason the Eurasians got a head start on technology is because they got a head start on agriculture, and they got that because among the wild plants and animals in their vicinity ten thousand years ago were ideal candidates for domestication. There were grasses that were genetically close to superplants that could be arrived at more or less by accident, just a few mutations away from big-head, nutritious grains, and animals that because of their social nature were genetically close to herdable animals that bred easily in captivity. (Maize in the Western Hemisphere took longer to domesticate in part because it had a greater genetic distance to travel away from its wild precursor.) And, of course, the key portion of the selection events that covered this ground, before modern agronomy, was what Darwin called "unconscious selection"—the largely unwitting and certainly uninformed bias implicit in the behavior patterns of people who had only the narrowest vision of what they were doing and why. Accidents of biogeography, and hence of environment, were the major causes, the constraints that "fixed" the opportunities of people wherever they lived. Thanks to living for millennia in close proximity to their many varieties of domesticated animals, Eurasians developed immunity to the various disease pathogens that jumped from their animal hosts to human hosts—here *is* a profound role played by human genes, and one confirmed beyond a shadow of a doubt—and when thanks to their technology they were able to travel long distances and encounter other peoples, their germs did many times the damage that their guns and steel did.

What are we to say about Diamond and his thesis? Is he a dread genetic determinist, or a dread environmental determinist? He is neither, of course, for both these species of bogeyman are as mythical as werewolves. By increasing the information we have about the various causes of the constraints that limit our current opportunities, he has increased our powers to avoid what we want to avoid, prevent what we want to prevent. Knowledge of the roles of our genes, and the genes of the other species around us, is not an enemy of human freedom, but one of its best friends.

Degrees of Freedom and the Search for Truth

The "decisions" made by *lineages* (of parasitic cells or of redwood trees, for instance) can be seen only by squinting just right. You have to adopt the intentional stance toward these curious ensembles of stuff, put time on *fast forward,* and hunt for the higher-level patterns to emerge from the mountains of data, which they do, with gratifying predictability. The more recognizable sort of decisions, made in real time by compact, salient individuals, had to await the birth of locomotion. Yes, trees can "decide" that spring has come and it is time to push out the blossoms, and clams can "decide" to clam up tight when they feel an alarming bump on their shells, but these options are so rudimentary, so close to being simple switches, that they are decisions by courtesy only. But even a simple switch, turned on and off by some environmental change, marks a *degree of freedom,* as the engineers say, and hence is something that needs to be controlled, one way or another. A system has a degree of freedom when there is an ensemble of possibilities of one kind or another, and which of these possibilities is actual at any time depends on whatever function or switch controls this degree of freedom. Switches (either on/off or multiple-choice) can be linked to each other in series, in parallel, and in arrays that combine both sorts of links. As arrays proliferate, forming larger switching networks, the degrees of freedom multiply dizzyingly, and the issues of control grow complex and non-linear. Any lineage equipped with such an array confronts a problem: What information *ought* to modulate passage through this array of forking paths in a multi-dimensional space of possibilities? That is what a brain is for.

A brain, with its banks of sensory inputs and motor outputs, is a localized device for mining the past environment for information that can then be refined into the gold of good expectations about the future. These hard-won expectations can then be used to modulate your choices—better than your conspecifics can modulate *their* choices. Speed is of the essence, since the environment is always changing and teeming with competitors, but so is accuracy (since among the competitors' options are such tactics as camouflage), and so is thrift (since everything costs something and has to pay for itself in the long run). These conditions on evolution generate a set of trade-offs, with a premium on swift, high-fidelity, high-relevance sensory attention. The arms race in future-production guarantees that each species will ignore whatever it can afford to ignore in its environment, a risky policy that may blindside it in the future, when a heretofore bland variable in its environment suddenly takes on fatal relevance.

This higher-order prospect of an environment rich in unanticipatable but relevant novelties poses another trade-off: Will it pay this lineage to invest in *learning*? There is a substantial overhead cost: Machinery must be installed to permit the switching networks to be re-designable in real time, during the individual organism's own lifetime, so that it can adjust its control functions in response to new patterns it detects in the world. Recall Drescher's (1991) distinction between *situation-action machines* and *choice machines* mentioned in Chapter 2. Situation-action machines are a collection of relatively simple switches, each one embodying an environmental rule of sorts: *If you encounter condition C, do A.* They are cost-effective for relatively simple organisms whose behavior is innately specified. Choice machines have a different set of mechanisms, which embody predictions: *If you encounter condition C, doing A would result in outcome Z* (*with probability p*). They generate several or many such predictions, and then evaluate them (using whatever values they have, or have developed), and this arrangement is cost-effective for organisms that are designed to learn during their own lifetimes. An organism can have both sorts of machines in its kit, relying on the former for quick-and-dirty lifesaving choices and relying on the latter for serious thinking about the future—a rudimentary faculty of practical reasoning.

Such fancy learning machinery will pay for itself only if there are enough occasions for learning (and the learning tends to be in the direction of new good habits, not new bad habits, of course). How much is enough? That depends on the circumstances, but there is no question that often there is not enough. "Use it or lose it" is a motto with many applications in the animal world. For instance, the brains of domesticated animals are significantly smaller than the brains of their nearest wild kin, and this is not just a by-product of selection for large muscle mass in animals raised for food. Domesticated animals can afford to be stupid and still have lots of offspring, for they have in effect outsourced many of their cognitive subtasks to another species, us, on which they have become parasitic. Like the tapeworms that have "decided" to trust us to handle all their locomotion and food-finding tasks for them so that they can drastically simplify their nervous systems, which they no longer need, domesticated animals would be in tough shape without their human hosts to live off. They are not *endo*parasites, living inside us, but they are still parasites.

We have arrived in the vicinity of the freedom of the bird, which can fly wherever it wants. Why does it want to fly where it wants to fly? It has its reasons. Its reasons are embodied in the settings of all the switches in its brain, and are endorsed, over the long run, by its continuing survival. Mostly, the things it cares to gather information about are the things that matter the most to its immediate well-being. The more pressure its ancestors have recently been under from wily competitors, the more likely it is to carry an investment in expensive equipment for countering that family of threats. When sailors first arrived in their sailing ships at remote islands in the Pacific inhabited by birds whose ancestors had not seen a predator in many thousands of years, they found birds so incurious, so unafraid of the large moving things approaching them, that the sailors could swagger right up and grab them. These birds could fly perfectly well, but no stealth was needed to capture them. They could fly wherever they wanted, but they didn't have very astute wants; there were reasons in the offing that they didn't know enough to make their own. They had plenty of bare opportunities to save themselves, but they lacked the information needed to act on them. These species of birds are largely extinct now, of course.

The arms race of predator and prey, as well as the competition among conspecifics for mates, and for the means for mates—food, shelter, territory, local standing, etc.—has given our biosphere hundreds of millions of years of R&D across a broad spectrum of parallel processing in millions of species at a time. At this very moment, trillions of organisms on this planet are engaged in a game of hide-and-seek. But for them it's not just a game; it's a matter of life and death.

Getting it right, not making mistakes, matters to them—indeed nothing matters more—but they don't, as a rule, appreciate this. They are the beneficiaries of equipment exquisitely designed to get what matters right, but when their equipment malfunctions and gets matters wrong, they have no resources, as a rule, for noticing this, let alone deploring it. They soldier on, unwittingly. The difference between how things seem and how things really are is just as fatal a gap for them as it can be for us, but they are largely oblivious to it. The *recognition* of the difference between appearance and reality is a human discovery. A few other species—some primates, some cetaceans, maybe even some birds—show signs of appreciating the phenomenon of "false belief"—*getting it wrong.* They exhibit sensitivity to the errors of others, and perhaps even some sensitivity to their own errors as errors, but they lack the capacity for the reflection required to *dwell* on this possibility, and so they cannot use this sensitivity in the deliberate design of repairs or improvements of their own seeking gear or hiding gear. That sort of bridging of the gap between appearance and reality is a wrinkle that we human beings alone have mastered.

We are the species that discovered doubt. Is there enough food laid by for winter? Have I miscalculated? Is my mate cheating on me? Should we have moved south? Is it safe to enter this cave? Other creatures are often visibly agitated by their own uncertainties about just such questions, but because they cannot actually *ask themselves these questions,* they cannot articulate their predicaments for themselves or take steps to improve their grip on the truth. They are stuck in a world of appearances, making the best they can of how things seem and seldom, if ever, worrying about whether how things seem is how they truly are. We alone can be racked with doubt, and we alone have been provoked by that epistemic itch to seek a remedy: better truth-seeking methods. Wanting to keep better track of our food supplies, our territories, our families, our enemies, we discovered the benefits of talking it over with others, asking questions, passing on lore. We invented culture.

It is culture that provides the fulcrum from which we can leverage ourselves into new territory. Culture provides the vantage point from which we can see how to change the trajectories into the future that have been laid down by the blind explorations of our genes. As Richard Dawkins has said, "The important point is that there is no

foresightless genes

general reason for expecting genetic influences to be any more irreversible than environmental ones" (Dawkins 1982, p. 13). But in order to reverse any such influence, you have to be able to recognize and understand it. It is only we human beings who have the long-range knowledge capable of identifying and then avoiding the pitfalls on the paths projected by our foresightless genes. Shared knowledge is the key to our greater freedom from "genetic determinism."

We haven't got to Symphony Hall yet, but we're getting closer.

Chapter 5

The wisdom inherent in the design of multicellular life forms can best be understood by adopting the intentional stance toward the whole process of evolution. From this perspective we can discern the free-floating rationales of the cooperative "choices" in non-zero-sum games that have guided the evolutionary R&D process to ever more sophisticated rational agents, expanding the capacity of life-forms to recognize and act on opportunities. Turning our backs on the misguided bugbear of "genetic determinism," we can see how evolution by natural selection provides for greater and greater degrees of freedom, but this is still not the freedom of human agency.

Chapter 6

Human culture is neither a miracle nor a straightforward addition to the tool kit provided to us by our genes to enhance their own fitness. In order to understand how a person can be both a creation of and a creator of culture, we need to explore the multi-stage evolutionary process from which culture, and human sociality, have emerged.

Notes on Sources and Further Reading

There are more extended developments of the ideas in this chapter in *Darwin's Dangerous Idea* (Dennett 1995), from which some of the paragraphs in this chapter are taken. John Maynard Smith's *Games, Sex and Evolution* (1988; especially Chapters 21 and 22) is an excellent introductory account of game theory in evolution, as is the revised edition of Richard Dawkins's *The Selfish Gene* (1976). Brian Skyrms's *Evolution of the Social Contract* (1996) carries the exposition through more recent research. For an arresting overview of the trend explored in this chapter, see Robert Wright's *Nonzero: The Logic of Human Destiny* (2000).

Our understanding of the evolutionary processes described here, especially the conflicts between genes that can be described from the intentional stance, is growing at a rapid pace. Many of today's specific claims (such as the number of genes in the human genome) may well be rescinded tomorrow, but the skeleton of theory and evidence that holds evolutionary biology together is remarkably robust and resilient. An excellent, if difficult, book surveying the steps in the transition from the simplest life-forms to human societies is Maynard Smith and Eörs Szathmáry's *The Major Transitions in Evolution* (1995); an easier version is their 1999 book, *The Origins of Life: From the Birth of Life to the Origins of Language.* For an authoritative overview of the state of knowledge circa the end of 2000, see *Evolution: From Molecules to Ecosystems* edited by Andrés Moya and Enrique Font (forthcoming), for a series of surveys on such topics as the evolution of multicellularity, conflicts that can arise in spite of the largely shared fate of mitochondrial and nuclear genes, the cost-benefit trade-offs of symbiosis, and many other fascinating topics.

Drescher's distinction between situation-action machines and choice machines usefully clarifies (and partially cuts across) the distinction I have drawn between Skinnerian and Popperian creatures (Dennett 1975, 1995, 1996A).

Chapter 6

THE EVOLUTION
OF OPEN MINDS

Human beings are not just clever brutes, resourceful agents looking out for themselves in a dangerous world, and they are not just herd animals either, unwittingly huddling together for mutual benefit that they needn't understand. Our sociality is a multi-layered phenomenon, replete with reverberant phenomena involving mutual recognition (of recognition of recognition . . .) and hence opportunities galore for such distinctively human activities as promise-making and promise-breaking, veneration and slander, punishment and honor, deception and self-deception. It is this environmental complexity that drives our control systems, our minds, into their own many layers of complexity, so that we can cope with the world around us effectively—if we are normal. There are unfortunate human beings who for one reason or another cannot, and they must live among us in a reduced status, rather like pets, at best, cared for and respected, restrained if necessary, loved and loving in their own limited ways, but not full participants in the human social world, and, of course, lacking morally significant free will. The problematic boundaries between them and the rest of us, and the supremely difficult issues that arise when individuals are up for promotion or demotion, will be the topic of a later chapter, but in order to lay the groundwork, we need to consider further how these unique complexities of human society and psyche evolved.

How Cultural Symbionts Turn Primates into Persons

A spider conducts operations that resemble those of a weaver, and a bee puts to shame many an architect in the construction of her cells. But what distinguishes the worst architect from the best of bees is this—that the architect raises his structure in imagination before he erects it in reality.

—Karl Marx, *Capital*

Culture makes things easier—or possible at all. And some of its changes seem more nearly inexorable ("evolutionary") than others.

—John Maynard Smith, "Models of Cultural and Genetic Change"

In species that lay their eggs and leave, never to share an environment with their offspring, genes are almost the only pathway of vertical descent or inheritance. Almost, but not quite, as we can see in a simple example: Take a species of butterfly that normally lays eggs on the leaves of a particular kind of plant, and consider what can develop when one female happens to lay her eggs on some other kind of leaf by accident. It is likely that the gene that is (most) responsible for this egg-laying habit works by getting offspring to "imprint" on whatever kind of leaf they first observe on hatching. This aberrant butterfly's offspring will repeat her "mistake" and instinctively lay *their* eggs on leaves that resemble their birthplace leaf. If her mistake happens to have been a happy accident, her lineage may prosper while others perish: The new leaf preference will be an adaptation *with no genetic change at all*.

This example highlights the element of *deixis,* or "pointing," involved in the kind of reference employed in genetic recipes. The gene in the butterfly's offspring says, in effect: Lay your eggs on something that looks like *this* (and a little finger points blindly out, falling on whatever target is there when the organism "looks" where the finger is pointing). Once the principle is understood, one can see it operating everywhere, especially in the multifarious developmental processes that depend on "cell memory." The butterfly didn't just deposit DNA on that leaf; she deposited eggs, and those egg cells contain all the reading machinery and initial raw materials for following the DNA recipes. This reading machinery, too, contains crucial information needed to make the offspring phenotype, and it is not coded in the genes; the genes just "point to" the ingredients and tell the reading machinery, in effect: Use *this* and *that* to make and fold the next protein.[1] If we arrange to alter these elements in the immediate environment of the gene-reading process, we can produce a change in output (like the offspring's altered leaf-choosing habit) and if it happens to be—like that habit—one that guarantees that the same alteration will tend to recur in the next generation's gene-reading environment, we have produced a phenotypic mutation (a mutation in the product, the vehicle that confronts natural selection) without any mutation in the genotype (the recipe). Cooks know that subtle changes in the texture of flour and sugar in different countries can have a profound effect on how their favorite recipes come out. They follow the recipe to the letter, reaching for *the stuff that is called flour here,* and get an unfamiliar cake. But if the new cake is a good cake, its recipe may be copied and followed by many cooks, creating a lineage of cakes quite distinct from their ancestors and from their contemporary kin in the home country. (I trust that aficionados will note the parallels between this point and the Twin Earth industry in philosophy. Those who don't get this parenthesis may consider themselves fortunate to be in the dark.)

Mother Nature is not a "gene centrist." That is, the process of natural selection doesn't favor transmitting information via genes when the same information (roughly) can be just as reliably, and more cheaply, provided by some other regularity in the world. There are the regularities supplied by the laws of physics (gravity, etc.) and by the long-term stabilities of environment that can be safely "expected" to persevere (salinity of the ocean, composition of the atmosphere, colors of things that can be used as triggers . . .). Since these conditions are more or less constant, they can be tacitly *presupposed* by the genetic recipes and not "mentioned." (Note that cake mixes sold in boxes often prescribe a different baking temperature, or the addition of extra flour or water, for high-altitude cooking, an instance of variance that obliges the recipe to mention something it could otherwise thriftily be silent about.)

1. The genes do code information to guide the construction of the *next* generation's reading machinery, of course, and to stock that generation's kitchen with raw materials, but other sources can also contribute to that specification, as we have just seen.

Among the regularities that can be presupposed by the gene recipes are those that are transmitted from generation to generation by social learning. These are just more cases of expectable environmental regularities, but they take on further importance because of the possibility that they themselves can be subject to selectional pruning (unlike gravity, for instance). Once the informational path of transmission is established, and becomes "relied upon" *by the genes* to do some of the carrying, it becomes subject to design improvements of its own, just like the myriad refinements that have beefed up the processes of DNA coding, replication, editing, and transmission over the eons. Genetic changes that tend to prolong parent-offspring contact and interaction, for instance, can raise the reliability of these pathways of social learning by giving them more time to operate, and then attentional biases (watch Mom!) can evolve to tune the transmission further. The path becomes a road becomes a highway, an informational channel *designed* by natural selection to enhance R&D in the lineages that rely on it.

In species in which parents and offspring live together for a time, there is a broad avenue for such vertical but non-genetic transmission of useful information or "tradition," such as food and habitat preferences (Avital and Jablonka 2000). As we have seen, horizontal transmission of *genetically transmitted* design, the sharing of useful genes with organisms other than your offspring or parents, has also been around since the earliest days of evolution, and has played a critical role in many of the most brilliant advances made by evolution, but these appear to be happy accidents, not designed pathways for spreading designs. Horizontal transmission of non-genetic information is a much more recent innovation in multicellular life-forms equipped with perceptual systems (animals, in short). Nowhere are its powers more evident than in our species, but we are not alone in enjoying its benefits. Monkeys being studied on a Japanese island have famously learned by imitation or observation the trick of cleaning wheat thrown on the beach by throwing sandy handfuls of it into the sea and then scooping the floating grains from the surface, and there is reason to believe that the dam-building technologies passed on by adult beavers to their young may include a substantial measure of observation and learning-by-imitating, if not formal instruction. There are, as usual in biology, some nice intermediate examples to illuminate the contrasts. Mountain goats trample a network of best-route paths across their territory, bequeathing this usefully prepared environment, as tidy as any human road system, not only to their offspring and grandoffspring, but to all the creatures who move through the area. Is this *cultural* transmission? Yes and no. The preservation of the uniformity relied upon depends on repetition of actions by individual goats, who have to be able to see what the other goats are doing. Is that imitation? What exactly is being *replicated*? It is hard to say. *cultural transmission*

But there is one species, *Homo sapiens,* that has made cultural transmission its information superhighway, generating great ramifying families of families of families of cultural entities, and transforming its members by the culturally transmitted habit of vigorously installing as much culture as possible in the young, as soon as they can absorb it. This innovation in horizontal transmission is so revolutionary that the primates that are its hosts deserve a new name. We could call them *euprimates*—superprimates—if we wanted a technical term. Or we could use the vernacular and call them *persons.* A person is a hominid with an infected brain, host to millions of cultural symbionts, and the chief enablers of these are the symbiont systems known as languages.

Which came first, language or culture? Like most chicken-and-egg puzzles, this one seems paradoxical only when you look at it simplistically. It is true that full blown language can't flourish as an institution among the members of a species until there is a community of sorts, with norms, and traditions, and recognition of individuals, and mutually understood roles. So there is a case to be made for the claim that some sort of culture precedes—and must precede—language. Chimpanzee communities have norms and traditions (of sorts), and recognition of individuals, and mutually understood roles (of sorts) without language, and they also show some modest cultural transmission: traditions or "technologies" for cracking nuts, fishing for termites, sponging water out of hard-to-reach sources. They even have some proto-symbols; in at least one chimp community, the slyly lascivious stroking of a plucked blade of grass by a male apparently means, to an onlooking female, something like "Va-va-voom!" or "Would you like to come up and see my etchings?" There are differences in hand clasps during grooming rituals that seem to be culturally, not genetically, transmitted. Looking back in our own evolutionary history, there is evidence (still hotly debated) of hominid control of fire going back a million years, and this was surely a culturally transmitted practice (not genetically transmitted, like the nest-digging practices of digger wasps), and yet language may well be a much more recent innovation, with estimates ranging from hundreds of thousands to only tens of thousands of years ago.

Culture and cultural transmission can exist without language, and not just in us hominids, and in chimpanzees, our closest surviving relatives. But it is language that opens the floodgates of cultural transmission that set us apart from all other species. Elaborate linguistic culture has apparently evolved only once on this planet—so far. (Neanderthals probably had language, so at one time there may have been two language-using species sharing the planet, but if so, they probably both inherited it from their shared ancestor.) Why haven't other species discovered this magnificent suite of adaptations? The list of features unique to *Homo sapiens* is familiar: control of fire, agriculture (but don't forget the fungus-farming ants), complex tools, language, religion, war (but remember the ants), art, music, weeping, and laughter. . . . In which order did these specialties emerge, and why? The historical facts are remote in time, but not quite inert; they do leave fossil traces that can be studied today by anthropologists, archaeologists, evolutionary geneticists, linguists, and others. What binds together all the interpretations of the data and governs the ongoing debates is Darwinian thinking—and it is not just about genes. Sometimes it is not about genes at all. Language has evolved only once, but *languages* have been evolving ever since the first language-using group split up into subgroups, and although there have definitely been genetic responses to the advent of language (brains have evolved anatomically to make them better word-processors), it is very unlikely that *any* of the evolved differences between, say, Finnish and Chinese, or Navajo and Tagalog, are due to any of the faint genetic differences that can be discerned (using sophisticated statistical analysis) between the human populations who speak these languages as their native tongue. Any human infant can learn any human language it is exposed to with equal ease, so far as we know. So the evolution of languages is not directly about the evolution of genes, but it has still been governed by the Darwinian constraints: All R&D is costly, and every new design has to pay for itself one way or another. If grammatical complexity of one sort or another persists, for instance, it does so for a reason, since *everything* in the biosphere is up for renewal, revision, or cancellation, all the time. Customs and habits will go extinct just as certainly as species, unless something keeps them going. Elaborate innovations—of language or of other human practices—don't just happen; they happen for reasons.

The question is: whose reasons? The lawyers ask "*Cui bono?*"— who benefits? To answer this question properly we need to make a bold leap of the imagination—without any magic feather to help us. You will notice, when you leap, a noisy crowd of hysterical bystanders warning you not to do it, imploring you to turn your back on this dangerous idea. The topic we are about to broach has an unparalleled power to upset the guardians of tradition and turn up the volume, but not the accuracy, of their criticisms. We are about to consider the prospect of *memes,* cultural replicators parallel to genes, and many who have considered the prospect just hate it. Let's try to understand it, first, and see if it is really so hateful. I will do my best to render vivid the grounds for the hatred, so as not to be accused of sugarcoating a poisonous idea, starting right now.

We see an ant laboriously climbing up a stalk of grass. Why is it doing that? Why is that adaptive? What good accrues to the ant by doing that? That is the wrong question to ask. No good at all accrues to the ant. Is it just a fluke, then? In fact, that's exactly what it is: a fluke! The ant's brain has been invaded by a lancet fluke (*Dicrocoelium dendriticum*), one of a gang of tiny parasitic worms that need to get themselves into the intestines of a sheep or cow in order to reproduce. (Salmon swim upstream; these parasitic worms drive ants up grass stalks, to improve their chances of being ingested by a passing ruminant.) The benefit is not to the reproductive prospects of the ant but the reproductive prospects of the fluke.[2]

In *The Selfish Gene* (1976), Richard Dawkins pointed out that we can think of some cultural items—which he dubbed *memes*—as parasites, too. They use human brains (instead of sheep stomachs) as their temporary homes, and jump from brain to brain to reproduce. Like the lancet flukes, they have been getting better and better at

2. Strictly speaking, to the reproductive prospects of the fluke's genes (or the fluke's "group"'s genes), for as Sober and Wilson (1998) point out (p. 18) in their use of *D. dendriticum* as an example of altruistic behavior, the fluke that actually does the driving in the brain is a sort of kamikaze pilot, who dies without any chance of passing on its own genes, benefiting its (asexually reproduced) near-clones in other parts of the ant.

negotiating this elaborate cycle (because of all the competition between memes for limited places in brains) and, also like the lancet flukes, they don't need to have a clue about how or why they do this. They are ingeniously designed informational structures that unwittingly exploit thinkers, but they aren't themselves thinkers. They don't have nervous systems; they don't even have bodies, in the ordinary sense. They are actually more like a simple virus than a worm (Dawkins 1993), because they travel light, instead of making a big body to move around in. Basically, a virus is just a string of nucleic acid (a gene) with attitude. (It also has a protein overcoat of sorts; a *viroid* is an even more naked gene, lacking the overcoat.) Similarly, a meme is an information-packet with attitude—a recipe or instruction manual for doing something cultural.

Memes are thus analogous to genes. What is a meme made of? It is made of information, which can be carried in *any* physical medium. Genes, genetic recipes, are all written in the physical medium of DNA, using a single canonical language, the alphabet of C, G, A, and T, triplets of which code for amino acids. Memes, cultural recipes, similarly depend on one physical medium or another for their continued existence (they aren't magic), but they can leap around from medium to medium, being translated from language to language, just like . . . recipes! Whether written in English in ink on paper, or spoken in Italian on videotape, or stored in a diagrammatic data structure on a computer's hard disk, the very same recipe for chocolate cake can be preserved, transmitted, copied. Since the proof of the pudding is in the eating, the likelihood of a recipe getting any of its physical copies replicated depends (mainly) on how successful the cake is. How successful the cake is at what? At getting a host to make another copy of the recipe and pass it on. *Cui bono?* Typically the eaters of the cake benefit, and that is why they treasure the recipe, making copies of it, and passing it on, but *whether or not* these "hosts" benefit, if one way or another the cake can encourage them to pass on the recipe, the recipe itself will benefit in the only way that matters for recipes: by being copied and thus prolonging its lineage. (We can imagine, for instance, that the recipe might be for making a cake that is, in fact, highly toxic but contains a powerful hallucinogen that gives people who eat it an overpowering, obsessive desire to make more copies of the recipe and share them with their friends.)

meme is beneficiary

In the domain of memes, the ultimate beneficiary, the beneficiary in terms of which the final cost-benefit calculations must apply, is: the meme itself, not its carriers. This is not to be heard as a bold empirical claim, ruling out (for instance) the role of individual human agents in devising, appreciating, and securing the spread and prolongation of cultural items. My claim is rather that we may adopt a perspective or point of view from which a wide variety of different empirical claims can be compared, including the traditional claims, and the evidence for them considered in a neutral setting, a setting that does not prejudice these questions. At first glance, this vision of culture may look more ominous than promising. If *this* is a kind of freedom, it is a strange kind indeed, it seems, and not in any way preferable to the bird's ignorant, if blissful, freedom to fly where it wants. In the analogy with the fluke, we are invited to consider a meme to be like a parasite that commandeers an organism for its own replicative benefit, but we should remember that such hitchhikers or symbionts can be classified into three fundamental categories: parasites, whose presence lowers the fitness of their host; commensals, whose presence is neutral (though, as the etymology reminds us, they "share the same table"); and mutualists, whose presence enhances the fitness of both host and guest. Since these varieties are arrayed along a continuum, the boundaries between them need not be too finely drawn; just where benefit drops to zero or turns to harm is not something to be directly measured by any practical test, though we can explore the consequences of these turning points in models. We should expect memes to come in all three varieties, too. Some memes surely enhance our fitness, making us more likely to have lots of descendants (e.g., methods of hygiene, child-rearing, food preparation); others are neutral—but may be good for us in other, more important regards (e.g., literacy, music, and art)—and some memes are surely deleterious to our genetic fitness, but even they may be good for us in other ways that matter more to us (the techniques of birth control are an obvious example). Trivially, the memes that persist will be those whose *own* fitness as replicators is greater, whatever their effects on our fitness, or indeed on our well-being in any sense. Thus it is a mistake to *assume* that the natural selection of a cultural trait is always "for cause"—always because of some perceived (or even misperceived) benefit it provides *to the host*. We can always ask if the hosts, the human agents that are the *vectors*, perceive some ben-

nat sel of cult. trait may not be "for cause"

human host may not perceive benefit / harm...

efit and (for that reason, good or bad) assist in the preservation and replication of the cultural item in question, but we must be prepared to entertain the answer that they do not. In other words, we must consider as a real possibility the hypothesis that the human hosts are, individually or as a group, either oblivious to, or agnostic about, or even positively dead set against some cultural item, which nevertheless is able to exploit its hosts as vectors. As George Williams has said,

> Within a society a meme may indeed enhance the happiness or fitness of its bearer, or it may not. If it can be horizontally transmitted at a greater rate than its bearer can reproduce, that bearer's fitness becomes largely irrelevant. The progress of cigarette smoking leaves a trail of corpses no less dead than those felled by a clone of spirochetes. (Williams 1988, p. 438)

There are many unanswered questions about memes, and many objections. Can the meme's-eye perspective be turned into a proper science of memetics, or is it "just" a vivid imagination-stretcher, a philosophical tool or toy, a metaphor that can't be made literal? It is too soon to tell. Most of the arguments that have been deployed against a science of memetics have been misguided and misinformed, and they betray a distinct whiff of disingenuousness or desperation. This is particularly evident when these arguments get repeated by people who manifestly don't understand them, since they faithfully and uncomprehendingly replicate minor errors that somehow got into the germ line! My favorite bad objection is the claim that cultural evolution is "Lamarckian," so it can't be "Darwinian," a mantra with several ill-considered variants, none of which hold water.[3] But it sounds good, doesn't it? It sounds like a sophisticated objection that must really hit those pesky ultra-Darwinians right where they live. *(Stop that crow!)* Pioneering research efforts now under way may mature into a substantial new discipline of

---

3. Briefly, Lamarckianism is the heresy of genetic transmission of acquired characteristics, but whose acquired characteristics—the memes' or their hosts'? Hosts pass on acquired parasites to their offspring all the time—no Lamarckian heresy there—and since memes have no germ-line/somatic-line distinction, there is no clear distinction between a mutation and an acquired characteristic of a meme. If "cultural evolution is Lamarckian" means either of these things, it is no objection to memetics; if it means something else, this has yet to emerge from the smoke screen.

memetics and prove these critics wrong. *(Eat that crow!)* Or they may not. There are still serious obstacles and objections that need to be met. (See the notes on further reading at the end of the chapter.) As I say, it is too soon to tell, but it doesn't matter for our purposes, because the main contribution from memes that we need on this occasion is, in fact, "just" philosophical or conceptual—and no less valuable for that: The meme's-eye perspective lets us *appreciate a possibility* that is otherwise very hard to take seriously. As we saw in Chapter 4 on libertarianism, there is a powerful conviction among many thinkers that *somehow* we have to be liberated from our brute biological heritage, if we are to have free will that matters morally. Since we can't engage in magical moral levitation, and we can't harness the quanta to carry us above our biology, we will have to look elsewhere for our liberation. Richard Dawkins closes *The Selfish Gene* with a ringing declaration:

> We have the power to defy the selfish genes of our birth and, if necessary, the selfish memes of our indoctrination. . . . We are built as gene machines and cultured as meme machines, but we have the power to turn against our creators. We, alone on earth, can rebel against the tyranny of the selfish replicators. (Dawkins 1976, p. 215)

But how can "we" do that? Dawkins doesn't say, but I think that the meme's-eye perspective, in fact, opens up just the prospects we need to fulfill his claim. It will take more than a few steps. The first is simply this: We can recognize that access to memes—good, bad, and indifferent—does have the effect of opening up a world of imagination to human beings that would otherwise be closed off. The salmon swimming upstream to spawn may be wily in a hundred ways, but she cannot even contemplate the prospect of abandoning her reproductive project and deciding instead to live out her days studying coastal geography or trying to learn Portuguese. The creation of a panoply of new *standpoints* is, to my mind, the most striking product of the euprimatic revolution. Whereas all other living things are designed by evolution to evaluate all options relative to the *summum bonum* of reproductive success, we can trade that quest for any of a thousand others as readily as a chameleon can change color. Birds and fish and even other mammals are quite immune to *fanaticism,* an affliction of cultural infection unique to our species, but, ironically, culture makes us susceptible to *open-minded about ends/means*

such pathologies by making us _open-minded_ about ends and means in a way no other animals are.

When an agent or intentional system makes a decision about which is the best course of action, all things considered, we need to know from whose perspective this optimality is being judged. A more or less default assumption, at least in the Western world, and especially among economists, is to treat the agent as a sort of punctate, Cartesian locus of well-being. What's in it for _me?_ Rational _self_-interest. But while there has to be something in the role of the self—something that defines the answer to the _Cui bono?_ question for the decision-maker under examination, there is no necessity in this default treatment, common as it is. A self-as-ultimate-beneficiary can in principle be indefinitely distributed. I can care for others or for a larger social structure, for instance. There is nothing that restricts me to a _me_ as contrasted to an _us._ (If you make yourself really small you can externalize virtually everything.) *not necessarily only self*

One tradition would speak here of "selfless" caring, but this creates more problems than it solves: The quest for "true" selflessness is a mission that is guaranteed to fail. It must fail not because we're no angels (we're no angels, but that's not the problem), but because the defining criteria of true selflessness are systematically elusive, as we shall see. It is better to think of the human capacity to rethink one's _summum bonum_ as the possibility of extending the domain of the self. I can still take my task to be looking out for Number One while including under Number One not just my own living body, but my family, the Chicago Bulls, Oxfam . . . you name it. Here is one good reason for treating the self this way: Suppose I am an agent in a bargaining situation, or in a Prisoner's Dilemma, or faced with a coercive offer, or an attempt at extortion. My problem is not resolved, or diminished, or even significantly adjusted, if the "self" I am protecting is other than my proper self, if I am not just trying to save my own skin, so to speak. An extortionist or a benefactor who knows what I care about is in a position to frame the situation to hit me where it matters to me, whatever matters to me.

We have arrived at the doors of Symphony Hall, but there is much more to be explored. We have to see _how_ cultural evolution, sometimes in harness with biological evolution, can produce the social conditions that compose the conceptual atmosphere, the air we breathe, when we conduct ourselves with the conviction that we are often free, *in a morally important sense,* to do whatever we decide.

The Diversity of Darwinian Explanations

Ethical ideas, political, religious, scientific ideas—all of these ideas and the institutions that embody them have arisen in very recent biological time, and not by magic. Culture didn't just descend on a band of hominids one day like a cloud of airborne germs. In order to understand how culture-borne ideas came to enlarge our selves, we have to look at the structure of the environment in which these ancestral agents must have acted. When we do this, we see a wide and largely unexplored variety of Darwinian hypotheses to test in our investigation of the history that has created our cultural heritage, and the reasons for the various parts of it.

When the cultural environment changes, a culture-borne habit can evaporate overnight, and this can send ripples back through the selective environment, so that there is a potent feedback cycle that speeds up evolution, often in directions we may come to regret. Consider a few examples. Walt Disney's cartoon feature *Bambi* was released in 1942 and changed American attitudes toward deer-hunting in the space of a few years (Cartmill 1993). Today the deer population in parts of the United States has become a serious public health problem, creating a minor epidemic of Lyme disease, spread by deer ticks who bite human beings who like to walk in wild country. In a single generation aluminum pots displaced the traditional Sukuma baskets of the *masonzo* culture along the shores of Lake Victoria in Africa:

> These watertight baskets were woven by women and used at celebrations as vessels for consuming vast quantities of *pombe,* a millet beer. . . . Blades of grass dyed with manganese were woven into the baskets in geometric patterns with a symbolic significance. It wasn't always possible to find out what the patterns meant because the arrival of the *mazabethi*—the aluminum dishes named after Queen Elizabeth that had been introduced on a large scale under British rule—had signified the end of the *masonzo* culture. I spoke to an old woman in a little village who, after more

than thirty years, was still incensed about the *mazabethi*. . . . "*Sisi wanawake,* we women, we used to weave baskets while sitting around and chatting with each other. I don't see anything wrong with that. Each woman did her best to make the most beautiful basket possible. The *mazabethi* put an end to all that." (Goldschmidt 1996, p. 39)

Even more sad is the effect reported of the introduction of steel axes to the Panare Indians of Venezuela.

> In the past, when stone axes were used, various individuals came together and worked communally to fell trees for a new garden. With the introduction of the steel ax, however, one man can clear a garden by himself . . . collaboration is no longer mandatory nor particularly frequent. (Milton 1992, pp. 37–42)

These people lost their traditional "web of cooperative interdependence," and now they are also losing a great deal of the knowledge they have amassed over centuries, of the fauna and flora of their own world. Often their very languages are extinguished, in a generation or two. Could something like that happen to us? Are there gifts from technology or science that could wreak as much havoc to our cultural milieu as these simple steel axes did to theirs? Why not? Our culture is made of the same sort of stuff as theirs. (*Stop that crow!*—only now, perhaps, we can all see that there actually might be good reasons to stop that crow.)

These examples show that culturally maintained features are highly volatile and easy to extinguish under some conditions, which is unsettling. But it is also hopeful. A cultural malignancy—such as a tradition of slavery or abuse of women—can sometimes be made to evaporate in as short a time, thanks to a few practical adjustments. Not all cultural features are so delicate. A culturally *enforced* habit may long outlive its usefulness, persisting thanks to sanctions imposed by the members of the culture, who may be oblivious to or only dimly appreciate the original rationale of their habit-turned-tradition. A taboo against eating pork, for instance, could have had an entirely sound rationale (free-floating or not) when it was first established, a rationale that lapsed long ago but is no longer required for the maintenance of the taboo. And if a feature is genetically anchored, the time lag between the cessation of its *raison d'être* and its extinction can be measured in hundreds of generations. Our sweet tooth, to take a well-worn example, made excellent sense back in our hunter-gatherer days, when energy-gathering was a matter of life or death. Now, with sugar ubiquitous in our environment, it is a curse that we must overcome with a variety of culture-borne countermeasures. (Hands up, all you genetic determinists who think this is impossible—hmm, I don't see any hands.)

There are numerous possibilities for complex interactions between genetic and cultural (and other environmental) factors. The differences in timescale alone ensure that. Consider, for instance, an incomplete survey of the possibilities for a Darwinian account of religion.[4] Religion is ubiquitous in human culture, and it flourishes in spite of its considerable costs. Any phenomenon that apparently exceeds the functional cries out for explanation. We don't marvel at a creature doggedly grubbing in the earth with its nose, for we figure it is seeking its food; if, however, it regularly interrupts its rooting with somersaults, we want to know why. What benefits are presumed (rightly or wrongly) to accrue to this excess activity? From an evolutionary point of view, religion appears to be a ubiquitous penchant for somersaults of the most elaborate sort, and as such it demands an explanation. There is no dearth of hypotheses. Religion (or some feature of religion) might be like:

Money: It is a well-designed cultural addition whose ubiquity can be readily explained and even justified: It's a Good Trick that one would expect to be rediscovered again and again, a case of convergent social evolution. The society benefits. (It is somewhat like the pheromone trails laid down by social insects to coordinate the activities of their fellows—its utility can be understood only in the context of the group, raising all the issues of group selection.)

A pyramid scheme: It is a cleverly designed con game passed on (culturally) through the generations of an elite, who use it to take advantage of their conspecifics. Only the elite benefit.

A pearl: It is the beautiful by-product of a rigid, genetically controlled mechanism responding to an unavoidable irritation; the organism thus protects itself from internal damage.

4. The next few paragraphs are drawn, with revisions, from Dennett 1997A.

A bowerbird's bower: It is the product of something analogous to runaway sexual selection, the elaboration of biological strategies caught on a positive feedback escalator.

Shivering: This apparently pointless agitation of the body actually has a benign role to play in maintaining the homeostatic balance, by raising the body temperature. The shiverer benefits, in most but not all circumstances in which it occurs.

Sneezing: Invading parasites have commandeered the organism and are driving it to destinations that benefit *them,* whatever its effects on the organism, like the fluke in the ant's brain.

The truth about religion might well be an amalgam of several of these hypotheses, or others. But even if this is so—especially if this is so—we will not get a clear vision of why religion exists until we have clearly distinguished these possibilities and put each of them to the test. They do not all pull in the same direction, but they are all instances of Darwinian thinking. All of the hypotheses seek to explain religion by uncovering some benefit, some work done to pay the costs, but they differ strikingly on the answer to *Cui bono?* Is it the group that benefits, or the elite, or the individual organism, or is it a "red queen effect" in which all parties have to run as fast as they can just to stay even, or is there yet some other evolutionary beneficiary? And none of these hypotheses invoke a "gene for religion"—though genes play a major role in setting up some of these possible preconditions for some aspects of religion.

There *may,* of course, actually be such things as genes for religion. For instance, heightened "religiosity" is a defining symptom of certain forms of epilepsy, and it is known that there are genetic predispositions for epilepsy. It *could* be that cultural environments—sets of traditions and practices and expectations—become amplifiers and shapers of certain rare phenotypes, tending to turn them into shamans or priests or prophets whose message is whatever the local message is (it's like learning your native tongue). In just such a way the "gift of prophecy" could actually "run in the family"—there would be a gene for it in exactly the same way there are genes for myopia or hypertension. (Yes, yes, I know; "strictly speaking" there *are* no such things as genes for myopia or hypertension; those so-called genes are only predispositions for those conditions. *Stop that crow!*) If there are any genes

evol of conditions that do amplifying

for religion, this is, in fact, one of the least interesting and least informative of the Darwinian possibilities. Much more important is the evolution (and maintenance, in the face of extinction) of the conditions
that might do the amplifying, and this is almost certainly not governed
by genes at all. It is cultural evolution.

While I'm fending off caricatures of Darwinian thinking, I
might as well sound an alert about another of them, which I will call the
nudist fallacy. *The American Sunbather* magazine (a few issues of which
came into my sweaty hands when I was a youngster) made a big deal, as
I recall, about the essential *naturalness* of nudity. It was a return to our
unclothed animal heritage, a way we could all get in touch with "the
way Mother Nature intended us to be." Nonsense. Not the part about
what Mother Nature intends—I am quite happy to defend the use of this
vivid phrase as shorthand for the free-floating rationales of the designs
evolution discovers and endorses. What is nonsense is the idea that what
Mother Nature intends is *ipso facto* good (for us now). By all means take
your clothes off whenever the spirit moves you, but don't make the mistake of supposing that by thus being "natural" you're improving your
condition in some way. (In fact, clothing is just as natural for our species
as a borrowed shell is for a hermit crab, who would be most unwise to
scurry around in the nude.) Myopia is natural, but thank goodness for
eyeglasses. Mother Nature intended us to eat all the sweet things we
could lay our hands on, but this is not a good reason for going with that
instinct. Many of the culturally evolved features of human life are quite
obviously cost-effective correctives for one superannuated "instinct" or
another (Campbell 1975)—and other features, as we shall see, are correctives to those correctives, and so forth. The Darwinian processes are
launched by the underlying competition among alleles in genomes, but
in our species the adaptations leave the launching pad far behind.

*not always
good now*

Nice Tools, but You Still Have to Use Them

Our opinions, gently nudged by circumstance, revise themselves under cover of inattention. We tell them, in a steady voice, No, I'm not interested in a change at present. But there is no stopping opinions. They don't care about whether we want to hold them or not; they do what they have to do.

—Nicholson Baker, *The Size of Thoughts*

In the past few decades, everyone has read or seen an endless number of books devoted to the culture of narcissism, of disbelief, of desire, or whatever. The argument in these books is always the same: what you imagine are your well-founded beliefs or preferences turn out to be nothing more than a set of reflexes implanted in you by the hidden assumptions of your "culture." You're not a skeptic about religion because you don't believe in the story of Noah and the Ark but because you are a member of the culture of disbelief.

—Adam Gopnik, *The New Yorker* (May 24, 1999)

One further source of resistance to Darwinian thinking in this charged context needs to be exposed and disarmed before we can proceed comfortably. A deep and persistent misunderstanding of Darwinian thinking is the idea that whenever we give an evolutionary explanation of a human phenomenon, in terms of either genes or memes, we must be denying that people think! This is sometimes a by-product of the caricature of genetic determinism, whose imaginary adherents say: "People don't think, they just have lots of unthinking instincts." But it can also be found in a caricature (sometimes, I must admit, a self-caricature) of theorists of cultural evolution who say, in effect: "My memes made me do it!"—as if memes (say, the memes of calculus or quantum physics) could do their work in their human hosts without requiring those human beings to do any thinking. Memes depend on human brains as their nesting places; human kidneys or lungs wouldn't do as alternative sites, because memes *depend on* the thinking powers of their hosts. *Being involved in thinking* is a meme's way of being put through its paces and tested by natural selection, just as *getting one's protein recipe followed and getting the result out in the world* is a gene's way of being tested. If memes are tools for thinking (and many of the best of them are just that), they have to be wielded for their phenotypic effects to show up. You still have to think.

It is true that a good Darwinian model of thinking will not look just like the traditional models. We do need to replace the bad old Cartesian model of a central, non-mechanical *res cogitans,* literally a *thinking thing,* that does the serious spiritual work. The Cartesian Theater, the imaginary place in the center of the brain "where it all comes together" for consciousness (and thinking) must be dismantled, and all the thinking work must be distributed to less fantastic agencies. In the next chapter, we will look in more detail at what follows from the fact that our thinking tasks get outsourced to semi-independent neural subcontractors in competition with each other, but the thinking still has to get done, and wherever thinking gets done, *people do things for reasons that are their reasons.*

So it is not a case of *memes versus reasons.* It is not even a case of *memes versus good reasons.* Explanations that purport to account for one thing or another by citing the reasoning done by thinking agents are not ruled out by a sound Darwinian approach. Far from it. The only position on reasons that memetics contradicts is the well-nigh incoherent position that supposes reasons somehow exist without support from biology at all, hanging from some Cartesian skyhook. A parody will expose the fallacy: "The people at Boeing are under the ludicrous misapprehension that they have *figured out* the design of their planes on sound scientific and engineering principles, and proven rigorously that the designs are as they should be, when *in fact* memetics shows us that all these design elements are simply the memes that have survived and spread among the social groups to which those airplane manufacturers belong." It is true, of course, that those memes have done well in those circles, but that does not compete with the good old-fashioned explanation in terms of well-planned, well-organized, well-conducted rational research and development. It supplements such an explanation.

Why would anyone think otherwise? Aside from occasional confusions on this score on the part of some would-be Darwinians, and aside from the caricatures, there is a more interesting reason. It has sometimes seemed that would-be memeticists deny any role to thinking because they occasionally mimic the perspective typically adopted by population geneticists, who deliberately ignore the actual operation of the phenotypes whose differential reproductive success determines the fate of the genes being studied. Population geneticists tend to shun all discussion of the bodies, structures, and real-world events that somehow compose selection events and instead just talk about the effects on the gene pool of one hypothesized change or another. It's as if lions and antelopes didn't actually lead lives, but just either procreated or not, depending on the fitness scores their bodies got. Imagine a tennis tournament in which the contestants just strip to the buff and get carefully examined, pairwise, by sports doctors and coaches who vote on which of each pair advances to the next round, until a winner is declared. Population geneticists would appreciate the point of such a strange practice, but would acknowledge that since the judges' criteria ought to be grounded in the rough-and-tumble of actual play, it is better to let the players go at it and let their actual contests decide the winners. Still, they would insist, you don't have to watch. Here is an expression of the standard rationale:

> As long as the proximate mechanisms result in heritable variation, adaptations will evolve by natural selection. There is a sense in which the specific proximate mechanism doesn't matter. If we select for long wings in fruit flies and get long wings, who cares about the specific developmental pathway? If the brainworm has evolved to sacrifice its life so that its group will end up in the liver of a cow, who cares how (or if) it thinks or feels as it burrows into the brain of the ant? (Sober and Wilson 1998, p. 193)

Similarly, the tussles among memes in brains *can* be ignored (after all, it's so messy and complicated) and we can stand back and just tabulate eventual winners and losers, but we mustn't forget that the contests do go on. Thinking happens, and how thinking happens affects which memes do well.

The Darwinian algorithms of evolution are *substrate-neutral*. They are not about proteins, or DNA, or even carbon-based life; they are about the effects of differential replication with mutation wherever it occurs, in whatever medium. This is especially important when we turn, as we are about to, to the evolution of morality. To appreciate this neutrality, consider a fantasy about another uniquely human creation, music. It is highly probable that we members of *H. sapiens* have some genetic predispositions vis-à-vis music. But whether or not this is probable, let's suppose it for the sake of a thought experiment. Let's suppose that our love of music, our responses to music, our talents for music, etc., are partially products of some genetically transmitted design features. And let's suppose that this distinguishes us from intelligent "Martians" (some non-human but culturally adept and communicative species), who utterly lack those human quirks in favor of music in their genetic birthright. A Martian research team visits our planet. One of their kind gets interested, in an intellectual way, in Earth-music, and endeavors to incorporate into its own perceptual capacities and proclivities all the discriminations, preferences, habits, etc., of a human music lover. While a normal human being has none of this work to do and is, in effect, a born music lover, for our imaginary Martian music is very definitely an acquired taste. But suppose the Martian does acquire it, by dint of diligent study and self-training. Now set aside the (ultimately boring) question of whether the Martian can *really* appreciate music "the way we human beings do." Consider, instead, the more interesting question of what the patterns are that distinguish great music from good music from so-so music from awful music.

What are the patterns the Martian is going to have to come to appreciate if it is to become a solid music critic, for instance? These are the patterns—deeply intertwined, surely, with the quirky genetic history of *H. sapiens* but independently describable—that a Darwinian theorist about music should most aspire to uncover. Suppose our Martian pioneer takes Earth-music back to Mars, and other Martians then take to this exotic new pastime and, following the lead of their pioneer, diligently imbue themselves with the requisite (but culture-borne) attitudes and dispositions. When *they* perform, enjoy, criticize the works of Mozart, the explanation of the source of their dispositions will be cultural, not genetic, but so what? It really doesn't matter (from some important points of view) whether someone is a "natural" (genetically designed) musician or an "artificial" (culturally designed) musician. The questions about the relations, the structures, the patterns that make this Mozart, or baroque music, or Earth-music will be substrate-neutral. And if, as seems likely, the Martian Hit Parade comes to include compositions that would never win an audience on Earth, the explanation of the differences in responsivity between Martians and Earthlings that account for these differences in taste will be neutral with regard to their genetic or cultural origins. Now if the Martians simply can't acquire these tastes, then they will never exhibit the

Summ. of forces of nature/nurture

patterns of preference and habit that could perpetuate the phenomenon; Martians just have tin ears, and music is not for them. But if they
can acquire the taste for music, it won't really matter how they acquire
it: The summation of the forces of nature and nurture in their development may arrive at the same sum by many different routes—all Darwinian. This thought experiment, science-fictional as it is, reminds us
of an important truth about differences between human musicians.
There are huge differences between those who have "natural" musical
talent and those who must inculcate it by internalizing large doses of
theory. It is, however, something approaching racism to declare that
only the former are true musicians, only the former *really* play music.
I suspect that eventually we will be able to identify genes "for" musical talent but music theory is, and should be, neutral with regard to
them.   *↙morality*
          So, too, should the theory that explains morality. It should be
neutral with regard to whether our moral attitudes, habits, preferences,
and proclivities are a product of genes or culture. It is an important
empirical question to what extent we are born "good natured," as de
Waal (1996) has said about chimpanzees, and to what extent we are
born "crooked" and have to be straightened out by culture, as Kant
has said about us: "*Aus so krummem Holze, als woraus der Mensch gemacht
ist, kann nichts ganz Gerades gezimmert werden*" (Out of timber so
crooked as that from which man is made nothing entirely straight can
be built). The explanation of how morality came to exist and why it
has the features it has will have to be Darwinian in either case. The
interplay between cultural and genetic transmission routes can be conducted only from a neutral perspective.

> Even groups that are genetically identical can differ profoundly at
> the phenotypic level because of cultural mechanisms, and these
> differences can be heritable in the only sense that matters as far as
> the process of natural selection is concerned. The fact that culture
> by itself can provide the ingredients required by the process of nat-
> ural selection gives culture the status that critics of biological deter-
> minism have emphasized. (Sober and Wilson 1998, p. 336)

          Explaining why music exists and why it has the properties it
has is one project, hardly begun. Explaining why morality exists and

why it has the properties it has is another project, on which somewhat more progress has been made, and this work will be the topic of the next chapter. Some of the guiding insights come from the work already discussed in Chapter 5 on evolutionary game theory. In recent years a growing multi-disciplinary band of researchers has been exploring the evolution of "cooperation," or "altruism," or "groupishness," or "virtue." Whether the results are called sociobiology or evolutionary psychology or Darwinian economics or political science or ethics naturalized or just an interesting branch of evolutionary biology, this approach describes a pattern that must be present in any such conflictual circumstance, whether it is embodied in genes or memes or some other cultural regularities. Several excellent books have recently appeared that survey and explain this research, and I will not attempt yet another primer when it has already been done so well by others (see the Notes on Sources and Further Reading at the end of the next chapter). Instead, I will step back and offer some interpretations to orient the work for our purposes, as well as some necessary correctives to a flood of misinterpretations that have dogged this research.

Chapter 6

A Darwinian approach to human culture permits us to sketch an explanatory path that can account for the major differences between us and our nearest animal relatives. Culture is a major innovation in evolutionary history. It provides one species, Homo sapiens, *with new topics to think about, new tools to think with, and—since the media of culture open up the possibility of cultural replicators whose own fitness is independent of our genetic fitness—new perspectives to think from.*

Chapter 7

The stability of the social conditions, individual practices, and attitudes that anchor our moral agency demands analysis and is beginning to receive it, from evolutionary theorists who recognize that culture itself must obey the constraints of evolution by natural selection. Contrary to the dire warnings of some critics, this approach does not subvert the ideals of morality; it provides much-needed support.

Notes on Sources and Further Reading

Animal Traditions (2000), by Eytan Avital and Eva Jablonka, is a fascinating investigation of the under-studied topic of animal tradition. See also my review (Dennett forthcoming B), which will appear in the *Journal of Evolutionary Biology,* and the review by Matteo Mameli, in *Biology and Philosophy,* 17:1 (2002).

Those who want to know more about Twin Earth can consult Andrew Pessin and Sanford Goldberg's anthology, *The Twin Earth Chronicles* (1996), or my essay "Beyond Belief" in *The Intentional Stance* (Dennett 1987).

On memes, see Blackmore 1999; Aunger, 2000, 2002; Dennett forthcoming C; and a special issue of *The Monist* on the epidemiology of ideas (Sperber 2001). In addition to *Darwin's Dangerous Idea* (Dennett 1995) and my essays in Aunger 2000 and Sperber 2001, I have written elsewhere on memes, in "The Evolution of Evaluators" (Dennett 2001); a review of Walter Burkert's *Creation of the Sacred:Tracks of Biology in Early Religions* (Dennett 1997A); and an overview essay, "The New Replicators," in *Encyclopedia of Evolution,* M. Pagels, ed. (Dennett 2002A).

An excellent examination of the question of why religions exist is Pascal Boyer's *Religion Explained:The Evolutionary Origins of Religious Thought* (2001).

An excellent article on using cladistic methods for the analysis of linguistic evolution is Gray and Jordan 2000, on Pacific language spread. Mark Ridley (1995, p. 258) has an account of lancet flukes, and a more detailed discussion occurs in Sober and Wilson 1998. Cloak 1975 converged with Dawkins 1976, on the *Cui bono?* question for cultural items: "The survival value of a cultural instruction is the same as its function; it is its value for the survival/replication of itself or its replica."

For a discussion of the error of pitting Darwinian explanation against reasons, see my comment on "A Critique of Evolutionary Archaeology," by James L. Boone and Eric Alden Smith, in *Current Anthropology* (Dennett 1998B).

Chapter 7

THE EVOLUTION OF MORAL AGENCY

I account for morality as an accidental capability produced, in its boundless stupidity, by a biological process that is normally opposed to the expression of such a capability.

—George Williams, in *Zygon*

If communities of genes and cells can evolve a system of rules that allow them to function as adaptive units, then why can't communities of individuals do the same? If they do, then groups will be like individuals, which is the proposition that we are seeking to establish.

—Elliott Sober and David Sloan Wilson, *Unto Others*

Is nature individualistic or communal? It is commonly thought—especially by those who fear any invocation of evolutionary considerations in ethics—that since Darwinism sees "nature red in tooth and claw," it can only subvert or discredit our ethical aspirations, never support them with new insights, new foundations. This is simply not true.

Benselfishness

We must indeed all hang together, or, most assuredly, we shall all hang separately.

—Benjamin Franklin to John Hancock, at the signing of the Declaration of Independence, July 4, 1776

This exhortation by Ben Franklin comes down to us through the ages, rippling red-white-and-blue in the breeze, redolent with the aroma of apple pie, a fine, noble, inspirational thing for our hero to have said, right? But wait a minute. Wasn't sly old Ben actually appealing to the craven, self-interested prudence of his listeners? Wise up, you cowards, and let me draw your attention to your actual predicament: Join or die. Which was it, a call for altruism and self-sacrifice or an appeal to those who knew which side their bread was buttered on? I propose that we concede that it was not, after all, a plea for *genuine* altruism (we'll consider later just what that might be, and whether it exists in significant amounts), but rather the expression of something still quite wonderful: a plea for a particular variety of *farsighted* self-interest, a kind of prudence that tends to get overwhelmed in competition because evolution is famously shortsighted, demanding immediate payoffs for all its innovations. I propose to call this particular variety of farsighted cooperative behavior *benselfishness,* in honor of Ben, but also suggestive of the fact that while this is a kind of selfishness, it's a *good* kind of selfishness. Were it not for the serendipity of Franklin's eloquence, I might have called it *euselfishness.*

Genuine, or pure, altruism is an elusive concept, an ideal that always seems to evaporate just when you get in position to reach out to grab it. It isn't clear what would count as genuine altruism, and paradox hovers constantly nearby. Imagine a world in which there is only one altruist and everybody else is selfish. The altruist and a selfish guy are stuck on an island with a rowboat that has room for only one. What should the altruist do? Should he volunteer to perish on the island, or is it better—more altruistic—for him to commandeer the rowboat, leaving the selfish guy to fend for himself, so that he can go help several selfish folks back on the mainland? An altruist shouldn't stupidly sacrifice himself for no gain—that's just being stupid. How crafty can an altruist be in exploiting others in order to achieve his altruistic ends? Consider the statutory safety briefing to passengers on airplanes. If you are traveling with a child, when the oxygen masks descend, first put on your own mask, then tend to your child. It seems that a parent can follow this advice with a clear conscience since it is probable (nothing is certain in life) that by taking care of yourself first, you will be better able to take care of your child, and your child's welfare is what matters most to you. That makes you an altruist. According to Elliott Sober and David Sloan Wilson, in their book *Unto Others: The Evolution and Psychology of Unselfish Behavior,* "The thesis of altruism, as we understand it, says that some people at least some of the time have the welfare of others as ends in themselves" (Sober and Wilson 1998, p. 228). Of course, it all depends on what counts as an end in itself. If you, selfish daydreamer that you are, prefer savoring in your imagination the future prospects of your child—if you prefer this activity to all others, and will take whatever steps are necessary to preserve the credibility of these parental flights of fancy by protecting your child, then you are no different from the miser who risks death to save his treasure chest from sinking to the bottom of the sea. If you have made the mistake of trading in your altruistic concern for your child for a selfish concern for your own peace of mind as you reflect on how you are sacrificing everything for your child, you are no true altruist. You're just taking all these steps in order to feel good about yourself.

And so forth, an all too familiar spiral of defeating conditions that we dutifully explore in intro philosophy class every year. It starts when we consider Socrates' notorious claim (in the *Meno*) that nobody ever desires evil for himself, a doctrine that is obviously false until it is shored up by adding that nobody ever *knowingly* desires something that is, *all things considered,* evil for himself. Is even that adjusted version true? Is it *impossible* or just *highly unlikely?* Is it just that anybody who did knowingly desire courses of action that were, all things considered, evil for himself would probably not last long enough to have offspring?

Mules are sterile because of their parents' genes, but not because they inherited "the gene for sterility" from their parents, for there is no such gene.[1] Sterility is a cul-de-sac, the end of a lineage, not something that can be passed on. Is an altruist rather like a mule, a more or less chance coming together of features that is perfectly *possible* but systematically unlikely to perpetuate itself? We should bear in

1. Mules have donkey fathers and horse mothers (usually—mules with donkey mothers are called hinnies); donkeys have 62 chromosomes, horses have 64 chromosomes (32 pairs), and mules have 63 unpairable chromosomes. There are very rare cases of fertile mules. And there are conditions under which there *could* be a sort of sterility gene. For instance, there could be a gene that in single dose (heterozygosity—a copy from either mother or father but not from both) provided a large benefit, so large that it persisted in spite of the fact that those with double doses of the gene (homozygotes) were sterile. This is a self-limiting possibility, since as the proportion of those with a single copy of the gene grows, the likelihood of both parents having a single copy, and both passing it on to their offspring, grows, and hence the proportion of sterile offspring grows, but they are a sink for the gene. The best-known instance of this quite familiar phenomenon, heterozygote superiority, is the resistance to malaria provided by a single dose of a gene that in double dose causes sickle cell anemia.

mind that although mules have no offspring, mules do proliferate at some times and places, thanks to indirect effects involving other species (such as the members of *Homo sapiens* who are also members of the British Mule Society, from which I got some of these details about mules). In fact, there are many ways in which evolution can sustain populations of organisms that seem, at first glance, to be systematically ruled out. There are conditions under which being altruistic—at least benselfish—is neither a genetic nor a cultural cul-de-sac, and these conditions have been exposed and clarified in a growing family of theoretical models.

The array of evolutionary game theoretic models that has been developed over the last few decades can be organized, with only a little rough shoehorning, into something like a genealogical tree of models, starting from an original seed, which has offspring, which have offspring, which have offspring, and so forth, and this tree exhibits—approximately—two interlocked trends: Parent models are simpler than their offspring, the next generation of models, and this increasing complexity of the models doesn't just bring increasing realism (with the models reflecting more and more of the actual complexities of the real world) but also increasing optimism! In the simplest models, altruism appears doomed. Aside from occasional short-lived freaks of nature, altruists seem to be ruled out by the fundamental principles of evolutionary theory, as impossible as perpetual motion machines. It's a dog-eat-dog world, and nice guys *inevitably* finish last. Then, as we add a few realistic touches, something in the direction of altruism appears and flourishes under certain conditions, and adding yet more layers of complexity seems to yield more varieties of quasi-altruism, pseudo-altruism, or whatever you want to call it. (I want to call it benselfishness.) Perhaps, it seems, as our models and theories get still closer to the complexity of the actual world, we will eventually arrive at genuine altruism, as a real possibility in the real world. Is this optimistic prospect an illusion? Is this bottom-up project as hopeless as trying to build a tower to the moon? You can't get there from here, say the anti-Darwinian skeptics. Don't even try. Or are the skeptics the ones who are confused, holding out for an inflated vision of altruism that is inaccessible by this bottom-up route only because it is inflated—a skyhook held aloft by hot air?

In any case, all the models show when and how benselfishness can flourish, and none of the models yet devised distinguishes between benselfishness and "genuine" altruism—if such a thing can be characterized. They all show conditions under which, bucking the constant headwind of evolution's myopia, organisms can come to be *designed* by evolution to cooperate, or more precisely designed to behave in such a way as to prefer the long-term welfare of the group to their immediate individual welfare.

The seed of this tree of models begins with the problem illustrated by the Prisoner's Dilemma. In these models, defection plays a role rather like the role of the Second Law of Thermodynamics in physics. Physicists are forever reminding us that things break down, things get muddled, things *don't* tend to repair themselves unless something special—such as a living thing, a local entropy-battler—intervenes. Economists, similarly, are forever reminding us that there is no such thing as a free lunch. Evolutionists in the same spirit remind us that freeloaders will always show up eventually, and when they do, they will soon enough win the local breeding contests unless something is put in place to prevent it. Whatever the local game, and whatever the costs and benefits to the *group* (the locally interacting population that must share the space and resources and risks), if it is possible to share the benefits of group action without paying one's share of the costs (one's dues, one might say), then those who pursue this selfish path will do better than those who don't. It's as simple as subtraction: *Net benefits* (benefits minus dues) have to be less than *gross benefits*, which is what the freeloader enjoys, by definition. All this must be true unless there are preventing conditions of one kind or another. Start with a uniform population of happy cooperators (they all have the cooperator gene, to keep it simple). They normally breed true, we may suppose, but what happens if a freeloader mutant appears in one generation of offspring? The freeloader does at least as well as the cooperators (since he doesn't pay his dues) and hence has a greater than average number of freeloader offspring. Pretty soon there is a growing tribe of freeloaders, and no matter how well or ill the group as a whole does (it probably does worse, weighed down as it is with all the freeloaders), within the group nobody does better than the freeloaders, who gradually come to dominate the group.

Of course, something may intervene to prevent this sad deterioration. You can imagine, if you like, that freeloaders tend to be ster-

blocking freeloaders.

ile, or infanticidal. What a lucky break for cooperators! You might as well imagine that Zeus likes to throw lightning bolts at freeloaders, keeping their numbers down (thank goodness) by his sport. Setting wishful fantasy aside, you can ask what might *evolve naturally* that would have the systematic effect of blocking the takeover by freeloaders, which must be assumed to be the default trend. As we have seen, this problem arose in the earliest days of life on this planet, in the intragenomic conflict between good genes and freeloading parasitic genes, and was solved by the evolution of counteracting mechanisms that could keep the freeloaders in check. Problems at that early and submicroscopic level were invisible to Darwin, of course, but he himself recognized the problem in the case of the social insects, whose extreme devotion to the group was a major challenge to the theory of natural selection. William Hamilton showed in his famous papers on "kin selection" how the social insects (and other highly social species) could evolve such patterns of cooperative instinct, and Richard Dawkins recast Hamilton's model into the perspective of the selfish gene. We are obliged to descend to the level of the gene to find the answer to the *Cui bono?* question in the extreme case of such self-sacrificial behavior because, as Sterelny and Griffiths vividly put it, "Perhaps a robin is being canny in choosing not to lay all the eggs she can, but a bee that stings an intruder at the certain cost of her own life cannot be saving anything for a rainy day" (Sterelny and Griffiths 1999, p. 157).

The pioneer models supposed, for simplicity, a single gene for "cooperate" and an alternative gene for "defect," and these genes were deemed to operate deterministically *at the biological level of behavior.* (Remember: This has nothing to do with the determinism or indeterminism of physics and everything to do with *design*. In these models, the individual organisms are stipulated to be old dogs that can't learn new tricks and are stuck as lifetime cooperators or defectors.) This is not much of an oversimplification if you're dealing with insects, whose behavioral routines are relatively rigid and tropistic (or *sphexish,* to use the term Douglas Hofstadter coined, in honor of the *Sphex* wasp), though even social insects can be strikingly facultative under some conditions, changing almost overnight from drone to worker when conditions in the colony demand redeployment, for instance.

These models show that defectors tend to do very well indeed, though they can pollute their own nests: As the proportion of freeloaders increases, they tend to meet each other more often, in costly bouts of mutual defection, and there aren't enough exploitable cooperators around to make up the difference. So the cooperators start making a comeback, but only until there are enough of them around to be worth preying on, at which point the freeloaders begin thriving again. But the models also exhibited some strange effects, settling into equilibria that don't match our expectations, and thus raising the prospect that at least some of the behavior of the models was artifactual, an unintended by-product of the oversimplifications rather than a reflection of something in the real world. (See Skyrms 1996 for a lucid treatment.) This is rather like the mythic discovery that, according to your aerodynamic model, bumblebees can't fly. Something must be wrong with your model, since there goes an airborne bumblebee. The model must be too simple, must be leaving out a complication that is actually a key to the bumblebee's manifest success. One simplification of these evolutionary game theory models was their super-abstractness. Individuals were just members of a set, drawn in random pairs for interactions that then determined their fate in the next phase, with no concern for their relative spatial locations in some world. It is as if the individual organisms lived on the Internet, as likely to interact with somebody halfway around the world as next door. (Actually, of course, interaccessibility on the Internet is highly ordered; some people are much "farther away"—harder to get to—than others, so these models would seriously oversimplify even the "global village" of the World Wide Web.) A second wave of models imposed a simplified spatiality, by adjusting the likelihood of encounters by a "viscosity" factor (the higher the viscosity of the imaginary space, the more likely you are to interact with somebody whose address is close to yours), and this simple change ushered in new opportunities for the evolution of cooperation, while also wiping out the embarrassing equilibria. It turns out that *neighborhood* makes a big difference. (Encroachment is what makes life interesting.) Neighborhood makes it more probable that you will interact with your own kind, so you get a better average payoff from any cooperative behaviors you engage in, since they are more likely to be reciprocated.

Then if we make the individual agents a little bit more sophisticated, allowing them some *choice* in who they interact with (just allowing them to refuse to play under some conditions, for starters), the simple space they all inhabit (not unlike the plane of the Life world) begins to pick up some structure: Clusters of like-acting agents begin to self-assemble, forming groups with different characters. Cooperators tend to find other cooperators, and defectors tend to get stuck having to associate with other defectors. This is all very suggestive, of course, but we are still a long way from altruism. For instance, wouldn't genuine altruists shun the selfish policy of finding like-minded altruists with whom to hang out? Wouldn't a genuine altruist go out of his way to be the lone altruist in a selfish group? That's where he's needed the most, it seems, not living it up among his fellow altruists. How merely benselfish of him! Besides, the agents in these models are still deemed to be pretty simpleminded old dogs, situation-action machines with a few preset switches that determine their "choices" in any encounter by the application of a simple rule. A vivid reminder of how simple the agents in these models are is that the tactics of self-segregation and ostracism that emerge from these models were already exploited at the macromolecular level of intragenomic conflict during the prokaryotic era. A model that needn't distinguish between a macromolecule and an adult human citizen is breathtakingly abstract.

When we make the agents still more facultative, more plastic, giving them the possibility of learning from their experience, adjusting the rules they were born with as a function of the encounters they have already had, things get more interesting still. The inevitability—note the term—of a group being swamped by freeloaders always depended on the assumption that everybody would be oblivious; there would be no capacity of the various individuals to notice what was happening, to raise the alarm, to deplore, to propose sanctions, to form vigilante groups, to brand or punish the freeloaders among them. Once we add simple versions of this reactivity, it ushers in a wave of new complexities. Dire conditions that had seemed inevitable now turn out to be preventable after all, thanks to the timely and well-aimed use of information by group members. The benselfish types now have a reason to punish too-pure "altruists"—the chumps or wimps who always let the freeloaders exploit them—since these pushovers help freeloaders flourish. So any mutations that permit the benselfish to distinguish themselves from pushovers will be favored, but then any freeloaders or pushovers who can disguise themselves as benselfish will tend to thrive, until the next phase of the arms race. A group's evolution of the capac-

distng btw benselfish + pushover.

→group conformism

ity for policing its members, by adopting the disposition among its members to punish violators (of whatever its other policies are), opens the floodgates to the *social* or *cultural* evolution of all manner of local norms. In a classic paper on cultural evolution, Rob Boyd and Peter Richerson show that if the cost of punishing is *relatively* low—something that can be virtually guaranteed whenever there emerges a practice of punishing those who don't punish—this creates an engine of group conformism of apparently unlimited scope and power. The title of the paper says it all: "Punishment Allows the Evolution of Cooperation (or Anything Else) in Sizable Groups" (Boyd and Richerson 1992).

So far, then, our evolutionary story has suggested the sorts of conditions that could have brought us, without skyhooks or other miracles, to a prudent disposition for cooperation, reinforced by the disposition we share with our fellow citizens to "punish" those who don't cooperate, but it is still a cold, robotic sort of mutually enforced nonaggression. As Allan Gibbard says,

> Human natural propensities were shaped by something it would be foolish to value in itself, namely multiplying one's own genes among later generations. Still, the kinds of coordination that helped our ancestors pass down their genes to form us are worth wanting—for better reasons. Darwinian forces shaped the concerns and feelings we know, and some of these are broadly moral. (Gibbard 1990, p. 327)

Broadly moral, but not purely moral. There is no sign yet of treating the welfare of others as an end in itself, for instance. This is probably as it should be, since we have yet to include anything distinctively human in the models, and one of our fairly comfortable initial intuitions about morality is that although non-human animals may be "good natured," as Frans de Waal says, they are not yet "the moral animal," as Robert Wright says. Still, since this sort of self-maintaining societal structure can now be seen to be a necessary precondition for the long-term flourishing of genuinely altruistic agents, it is reassuring to see how little must be presupposed to get it to evolve and to sustain itself: The very simplicity and relative rigidity of the abilities to discriminate the freeloaders from the good citizens, and the dispositions to "punish," show that as far as *this* feature of culture is concerned, it

self-maint societal structure

[handwritten: this culture could be before lang]

[handwritten margin: dis unn against other]

could predate language and convention and ceremony. We're not talking about trial by jury and public denunciation here; we're talking
about an unreflective, "brute" inclination to channel some risky
aggression against those of one's group one has discriminated as norm-
violators. It would be reasonable to look for evidence of this sort of
long-term maintenance of local "customs" among packs of wolves or
troops of monkeys or apes, for instance. Whether or not we find that
this station on the way to fully developed human culture is clearly
occupied by some other species, it provides a certain sort of relief from
skepticism: a *possible* Just So Story to get us gradually from animals that
are merely social in the manner of bees and ants to animals that have a
taste for cultural transmission and inculcation, disposed to attend to the
nuances of approval and disapproval, disposed to be enlisted in transi-
tory enforcement posses, disposed to prefer the comfort of acceptance
to the threat of group censure. And with this transition, groups become
effective repositories of recently discovered "knowledge," not having
to wait for the genetic evolution of each new Good Trick to spread to
fixation through the population, since it can be much more swiftly
spread by group conformism. A price well worth paying for access to
this brighter tempo of discovery is a certain vulnerability to something
like myth, local *mis*discoveries that nevertheless sell like hotcakes in the
structured conformism of the group.

Being Good in Order to Seem Good

Jesus is coming. Look busy!
—bumper sticker

Conscience is the inner voice that warns us that someone might be
looking.
—H. L. Mencken, *Prejudices*

The specter of defection hangs over us all, evolutionary original sin,
with its perennially tempting reflection: How can it *not* be rational to
defect here? If the other guy defects (or if "everybody does it"), then
you're a patsy if you don't defect as well, and if the other guy doesn't

defect, you make out like a bandit by defecting. And if everybody knows this, how can anybody ever cooperate? When payoffs are in the short term, how can evolution ignore them, and when we consider that life is short, how can we ourselves ignore these payoffs? Fear of punishment and a desire for acceptance will take us past the easy cases, by changing the expected payoff. As thinkers have recognized for centuries, it is not hard to see why it is rational to cooperate when Big Brother is watching. Any society that was lucky enough to harbor a belief in a vigilant, omnipresent God—who could be expected to mete out punishment in the afterlife more than compensating for any local gains—would be a society populated by citizens who could be counted on to do what that God commanded, even when out of sight of their fellow citizens. Note that for this myth to arise and flourish, there need not have been an intelligent author who understood this rationale, any more than there must have been an intelligent promulgator of the policies that evolved to secure compliance between potentially rival genes in meiosis. Human beings could be the unwitting beneficiaries of this group adaptation without anybody having figured out its free-floating rationale. But as critics since Nietzsche have insisted, a "morality" thus based on fear of God is neither as noble, nor as stable, as we would like. What would happen to a society in which this useful scaffolding began to break down, or never existed in the first place? Would there be no way for its members to evolve robust habits of cooperation?

What about the hard cases, in which one can be pretty sure one will not be detected cheating? In these cases the voice of temptation speaks with alarming rationality: *Nobody will ever know, and think of what you can gain!* When we enter the world in which decision-making has to deal with serious temptation, and with the unbounded terraces of reflection that can accompany our struggles with such temptation, we have left the free will of the birds behind and begun to explore the problematic territory of human free will, the only variety that carries moral weight. Tradition places the burden of all that moral weight on an imaginary functionary, the immortal, immaterial, miracle-working soul, but once we look more closely at the evolutionary antecedents of our human control systems, we can reverse-engineer that soul and see why some of its parts work the way they do.

According to Sallust, Cato was a noble man indeed: "*Esse quam videri bonus malebat*"—he preferred to be good rather than to seem so. If

Robert Frank is right, then Cato was one of those advanced souls who have managed to invert the policy that made us moral in the first place: *Malo esse bonus ut videar*—I prefer to be good *in order to seem* good. In *Passions within Reason: The Strategic Role of the Emotions,* Frank argues that the next plateau in the evolution of freedom is achieved when our ancestors first confronted and learned to solve what he calls *commitment problems.* A commitment problem "arises when it is in a person's interest to make a binding commitment to behave in a way that will later seem contrary to self-interest" (Frank 1988, p. 47). We have already encountered the basic structure of a commitment problem in the Prisoner's Dilemma: The evolutionary fate of cooperators and defectors is powerfully affected by the presence or absence of fake cooperators, or bluffers. This creates a selection pressure for bluff-detection and sets off an arms race of exposure and concealment of strategy. When the free-floating rationales of this competitive arena are captured within the flexible control systems of human agents, the tempo picks up, and the issue is transformed from the impersonal (which agents will do better under these conditions right now, cooperators or defectors?) to the personal (what should *I* do under these conditions, cooperate or defect?). When evolution gets around to creating agents that can learn, and reflect, and consider rationally what they ought to do next, it confronts these agents with a new version of the commitment problem: how to commit to something *and convince others you have done so.* Wearing a cap that says "I'm a cooperator" is not going to take you far in a world of other rational agents on the lookout for ploys. According to Frank, over evolutionary time we "learned" how to harness our emotions to the task of keeping us from being too rational, and—just as important—earning us a reputation for not being too rational. It is our unwanted excess of myopic or local rationality, Frank claims, that makes us so vulnerable to temptations and threats, vulnerable to "offers we can't refuse," as the Godfather says. Part of becoming a truly responsible agent, a good citizen, is making oneself into a being that can *be relied upon* to be relatively impervious to such offers.

First, why should you want to have such a reputation? Well, if you have the reputation, the Mafia will leave you alone, since they will calculate that their coercive offers probably won't work on you, so why waste a good horse's head? Even more important, your reputation will appeal to the choosiness of your fellow group members, who know all about the risks of being taken in by a defector, and who will scout around for somebody they think they can rely on to resist temptation. We noted in the previous section that cooperators tend to hang out with cooperators, and defectors with defectors. "Commitment problems abound and, if cooperators can find one another, material advantages are there for the taking," Frank observes (1988, p. 249), and the advantages of being a cooperator in a group of cooperators has been demonstrated in a host of evolutionary models. If you are fortunate enough to find yourself in a group of cooperators, is this just luck? Not if the group has an entrance exam. But are you then just lucky to have the talent for cooperation that lets you pass the exam? Perhaps, but being lucky to be talented is better than just being lucky. (I will have more to say about luck later.)

It is benselfish to want to have an impeccable reputation, but how on earth can you establish this? Since talk is cheap, anybody who is asked will swear on a stack of Bibles that they will never defect. Unless there is some other way to discern the cooperators among the defectors, there is scant chance of building stable groups of *rational* cooperators. (Remember: The somatic-line cooperators that compose most of your body are *ballistic* intentional systems, quite reliably robotic and impervious to temptation, but now we're talking about building not a body but a corporation of highly rational individuals, like the Boston Symphony Orchestra.) And for there to be a trustworthy signal of reliability, it must be, as Amotz Zahavi (1987) has shown us, a costly signal—something that cannot be cheaply faked. Swearing on a Bible is an empty ceremony that *cannot* convey usable information, since if it were to get started as a signal of reliability, it would immediately be copied and used by all the unreliable types, and hence lose its credibility and fall into disuse. You might try to save it by inflating the ceremony—I'll swear on *two* Bibles, I'll swear on a *stack* of Bibles—but the fruitlessness of this inflation is nicely alluded to in the idiom, our mythical paradigm of a failed attempt to demonstrate trustworthiness.[2] So here is the main problem: not just how can you make yourself into an agent that can be trusted in commitment problem cases, but how can you credibly advertise the fact that you are to be so trusted?

2. So why does the practice of taking an oath on the Bible persist? Because, quite independently *today* of the participant's belief in divine retribution, it signals one's deliberate entrance into the jeopardy of perjury, taking on the variable but still substantial risk of mundane retribution.

Sometimes one problem can be solved by another problem. This is especially true when the problem is confronted by Mother Nature, that master opportunist. We have a problem of self-control that is truly hard—costly—for us to solve. According to Frank, the fact that it is costly to solve is a blessing, not a curse. It is the problem exemplified by Ulysses and the Sirens, where the trick is to devise some way of tying yourself to the mast and blocking your sailors' ears with wax so that you can't act on your strongest inclination of the moment. (The trick is to arrange it so that "at time *t*" your will is *ineffective*.) Ulysses knows perfectly well the long-term benefits of adopting the policy of avoiding the Sirens when they sing their seductive song, but he also knows he is disposed in many circumstances to overvalue immediate payoffs, so he needs to protect himself from a somewhat misshapen preference structure that he expects will impose itself on him when time *t* rolls around. He knows himself, and he knows what evolution has provided for him: a slightly second-rate faculty of reason that will cause him to take the immediate payoff ("I couldn't do otherwise," he'll say, as he jumps into the Sirens' arms)—unless he takes steps now to distribute his decision-making over more favorable times and attitudes. His seduction by the Sirens is not *inevitable*, provided he has enough lead time to prepare his avoiding move. As Frank observes,

> It is important to stress that the experimental literature does not say that immediate payoffs get *too much* weight in every situation. It says only that they always get *very heavy* weight. On balance, that was likely a good thing in the environments in which we evolved. When selection pressures are intense, current payoffs are often the *only* ones that matter. The present, after all, is the gateway to the future. (Frank 1988, p. 89)

Ulysses' problem is not a moral problem; it is a prudential problem, of the sort that can afflict the most selfish, least altruistic of agents. To the selfish agent, it is the problem of how to avoid falling for short-term selfish gains at the expense of longer-term selfish gains, a problem of mastering himself for a life of greater prudential success. Before turning to Frank's account of how, by solving this prudential problem we carry ourselves all the way to morality, we need to look in a bit more detail at the problem of temptation.

Learning to Deal with Yourself

Intertemporal bargaining seems to be a rather artificial process unlikely to have arisen in lower animals. It was the human race that vastly expanded an individual's scope of choice and discovered that free choice often serves us worse than bald necessity.

—George Ainslie, *Breakdown of Will*

As the old-time Maine farmer began to hitch up his overalls after using the outhouse, a quarter rolled out of his pocket and fell down the hole. "Dang!" he said, and pulled a five-dollar bill out of his wallet and threw it down the hole after the quarter. "Why on earth did you do that?" he was asked. "You don't think I'm going down there for a quarter, do you?" he replied. Raising the stakes for ourselves changes the task of self-control we confront. We all tend to have problems with temptation that are nicely revealed in a few simple questions:

1. *Which would you prefer: a dollar right now or a dollar tomorrow?* If you are like most normal people you prefer the dollar now, for obvious reasons. The sooner you get it the sooner you can put it to use, and who knows what the future will hold? If, weirdly, you were entirely indifferent between the choice of a dollar now or tomorrow or next week or next year, we would say that you do *not discount the future*. It is obviously rational to discount the future, but how much?

2. *Which would you prefer: a dollar right now or a dollar-fifty tomorrow?* If you prefer the dollar-fifty tomorrow, how about a dollar and a quarter? How about a dollar and a dime? At some point we'll find a choice about which you are indifferent, and that will fix two points on a curve, your discount curve for the future. We might gather lots of data of this sort in order to plot lots of points on your particular curve, using money as a handy measuring system (standing in for a much wider set of your preferences: Which would you prefer, to be pain-free today or pain-free a week from today? Which would you prefer, fame tomorrow or fame next year?). Suppose you are indifferent regarding question 2. A dollar today or a dollar-fifty tomorrow strike you as equally desirable. Then consider the next question:

3. *Which would you prefer: a dollar next Tuesday or a dollar-fifty next Wednesday?* This is the same question as the previous one, but just seen from farther away in time. But you may well find that your answers

don't match. If you are like most people, a dollar right now is quite hard to turn down in favor of a dollar-fifty tomorrow, while it's relatively easy to do the prudent thing and sign up for the dollar-fifty next Wednesday instead of signing up for the dollar next Tuesday. If you tend to prefer a dollar now to a dollar-fifty tomorrow but also prefer a dollar-fifty on Wednesday to a dollar on Tuesday, you have a conflict; you will discover a shift in your preferences at some point in time between now and next Tuesday, a shift brought about by nothing but the passage of time.

Our susceptibility to these *intertemporal conflicts* is a glitch, a foible, an anomaly in our basic competence as decision-makers or choosers, and it lies at the heart of a remarkable theory of human will developed by the psychiatrist George Ainslie and recently given an accessible presentation in his book *Breakdown of Will* (2001). People may discount the future at different rates, and there is no right answer to how steeply one should discount the future, but whatever your rate, if you were rational about how you apply it, you would apply it so that no intertemporal conflicts arise: The cool-headed choice you make now for next year is the same choice you would make when next year rolls around. *Succumbing to temptation* is being deflected from your rational policy (whatever it is) in a way you would rationally like to avoid, if only you could. What shape should your discount curve take? Figure 7.1 shows two basic types of curve superimposed: the gradual *exponential* curve and the deeply bowed, steeply rising *hyperbolic* curve.

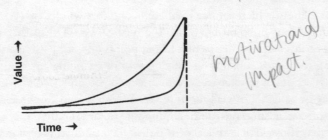

Figure 7.1 An exponential discount curve and a hyperbolic (more bowed) curve from the same reward. As time passes (rightward along the horizontal axis), the motivational impact—*the value*—of a subject's goals gets closer to its undiscounted size, which is depicted by the vertical line (Ainslie 2001, p. 31).

It can be shown (see, visually, in Figure 7.2) that an exponential discount rate can't produce these anomalies, but a hyperbolic discount rate (see Figure 7.3), by having a steep tail, can.

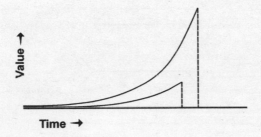

Figure 7.2   Conventional (exponential) discount curves from two rewards of different sizes, available at different times. At every point at which the subject might evaluate earlier and later rewards, their values stay proportional to their objective sizes (Ainslie 2001, p. 32).

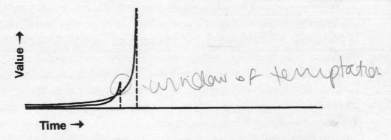

Figure 7.3   Hyperbolic discount curves from two rewards of different sizes available at different times. The smaller reward is temporarily preferred for a period before it is available, as shown by the portion of its curve that projects above that from the later, larger reward (Ainslie 2001, p. 32).

Where that snaggle-tooth hyperbolic hook of the smaller reward briefly crosses the curve of the greater reward is where your window of temptation is open: a brief period of time when the smaller reward *seems more valuable* than the greater reward. Voluminous testing under many conditions has shown that we, like other animals, are innately equipped with hyperbolic discount rates. "The human race evolved with a very regular but deeply bowed discount curve for evaluating the future" (Ainslie 2001, p. 46). This, Ainslie notes, is an illusion rather like the Müller-Lyer illusion.

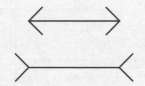

Figure 7.4   Müller-Lyer Illusion

We may know—thanks to measurement—that the two lines are the same length, but that doesn't stop the illusion from exerting a powerful force on us. We can learn to compensate for that naturally illusory outcome, overruling it with a deliberate, conscious correction. Similarly, utility theory (and measurement) can convince us that an exponential discount rate is right, and we can then learn to compensate for the hyperbolic discount rates we were born with. It's an unnatural act, but one well worth learning to perform. Some of us do it better than others.

The desirability of rationalizing our behavior along exponential lines is at least dimly appreciated by us, but how on earth do we do it? Where does the oomph come from to overrule our own instincts? Tradition would say it comes from some psychic force called *willpower,* but this just names the phenomenon and postpones explanation. How is "willpower" implemented in our brains? According to Ainslie, we get it from a competitive situation in which "interests" engage in what he calls "intertemporal bargaining." These "interests" are temporary agents of sorts, homunculi representing various reward possibilities:

> An agent who discounts reward hyperbolically is not the straightforward value estimator that an exponential discounter is supposed to be. Rather, it is a succession of estimators whose conclusions differ: as time elapses, these estimators shift their relationship with one another between cooperation on a common goal and competition for mutually exclusive goals. Ulysses plan-

ning for the Sirens must treat Ulysses hearing them as a separate person, to be influenced if possible and forestalled if not. (Ainslie 2001, p. 40)

The "power bargaining" engaged in by these "groups of reward-seeking processes" is a self-equilibrating process that needs "no ego or judge or other philosopher-king, no organ of unity or continuity, although it will predict how such an organ may appear to operate" (p. 62). As Ainslie describes this phenomenon, it is a competition for selection in which the competitors can co-opt and exploit each other, and it is none other (I suspect) than the opponent process of "striving will" sketchily imagined by Kane. It does, indeed, contribute importantly to the unpredictability of human choice, not by harnessing quantum randomness as Kane hoped, but by having built into it a recursive feature that systematically thwarts prediction: When we choose, *we reflectively use our choice as a predictor of what our choices in the future will be;* our very self-consciousness about our choices creates a recursive loop that renders our choices indefinitely sensitive to further considerations.

The orderly internal marketplace pictured by conventional utility theory becomes a complicated free-for-all, where to prevail an option not only has to promise more than its competitors, but also act strategically to keep the competitors from turning the tables later on. (Ainslie 2001, p. 40)

Ainslie analyzes how the microstrategies of these homunculi bundle rewards together, thereby creating an approximation of an exponential discount rate, generating "rules" and resolutions that in turn generate justifications for minor exemptions (it will be easier for me to keep to my diet if I'm not too strict on myself, so—since it's my birthday—I'll reward myself with a little bit of cake . . .), which in turn generate further moves and countermoves, a snowballing chaos of internal challenges. For instance: "Once I expect myself to find an exception whenever the urge is strong, I no longer have a credible prospect of the whole series of later rewards—the cumulative benefits of my diet—available to choose. In this way, hyperbolic discount curves make self-control a matter of self-prediction" (p. 87).

A recovering alcoholic may expect to resist taking a drink, but this expectation surprisingly disappoints her, and when she notices this she loses confidence in her expectation; if her expectation falls below being enough to stake against her thirst, her disappointment is apt to become a self-confirming prophecy. But if this prospect is itself daunting enough in the period before it becomes preferred, she'll look for other incentives to oppose her thirst before it becomes too strong, and thus raise her expectation of not drinking, and so on—all before she's actually taken a drink. Her choice is doubtless determined in advance, in the same sense that all events have strict causes that have causes in turn; but what immediately determines her choice is the interplay of elements that, even if well known in themselves, make the outcome unpredictable when they interact recursively.

Hyperbolic discounting makes decision making a crowd phenomenon, with the crowd made up of the successive dispositions to choose that the individual has over time. At each moment she makes the choice that looks best to her; but a big part of this picture is her expectation of how she'll choose at later times, an expectation that is mostly founded on how she has chosen at previous times. (Ainslie 2001, p. 131)

Ainslie's theory of the will generates explanations of more than a few phenomena that have baffled other theorists (or just been conveniently ignored by them), on such topics as addiction and compulsion, "premature satiation," self-deception and despair, "legalistic" thinking and spontaneity. The price one must pay for this theoretical fecundity is some initially counterintuitive premises: in particular, rewards and pleasures must be distinguished. Rewards are, by definition, "any experience that tends to cause repetition of the behavior it follows" and some such experiences are positively painful, however much they may enhance the replicative disposition (the intracerebral fitness, you might say) of that behavior. It is a difficult theory, bristling with novelties that require one to set aside dear old habits of thought, and I have only skimmed the most interesting conclusions off the top in this presentation. It has not yet received the attention it deserves, so just which of its many tempting conclusions deserve endorsement is an open question, but there is no question that it is a fine addition to the wealth of recent work that applies an evolutionary perspective to the traditional philosophical questions of will and mind. It even has some unsettling observations to make on the topic of the elusiveness of morality, and the ways in which our best-formulated rules can come to haunt us with unintended consequences, but those are topics for another occasion. We still haven't arrived at the moral arena, but Robert Frank proposes a path.

Our Costly Merit Badges

Suppose you place a candy in front of a small child and tell her that she may take it, but if she can wait fifteen minutes, she can have two. How good are children at this feat of delaying gratification? Not very. Children exhibit significant variation in this capacity for self-control, and whether these differences are due in the main to genetic differences or early childhood environmental differences or sheer chance, they are not *inevitable;* they can be diminished (or enhanced) by simple strategies of self-distraction, or the right kinds of concentration. (For instance, children can learn to hold out for the second piece of candy by concentrating on the delightful properties of something else that isn't available—nice crunchy, salty pretzels, for instance, or a favorite toy.) Some good strategies invoke cool reason and others invoke competing hot passion. These self-manipulation proposals, by the way, oppose an influential theme in moral philosophy, the theme attributed to Immanuel Kant that stresses the second-rate, ignoble nature of such laying-on of *merely emotional* crutches. The Kantian ideal is a fantasy in which you somehow strengthen your pure-reasoning muscle to such a fine pitch that you can make pure, emotionless judgments untainted by tawdry guilt feelings or base longings for love and acceptance. Kant held that such judgments are not only the best sort of moral judgments, they are the only sort of judgments that count as moral at all. Enlivening reflection with base appeals to emotion may be fine for training children, but the presence of those training wheels actually disqualifies their judgments for moral consideration. Is this perhaps a case in which holding out for perfection—a job-related disability in philosophers—conceals the best path?

According to Frank, the evolutionary beauty of this co-opting of emotion to play such a role in self-control is that it provides at the same time a basis for costly signaling of precisely this triumph: Others get to see that you are one of those emotional folks who can be counted on to care *passionately* about your commitments; it is not that you are *crazy* or *irrational* but that you put an irrationally high price (from the myopic perspective of the critic) on your integrity. You get to wear your heart on your sleeve, and a costly heart it is. The trick to gaining the reputation for being good, a valuable prize indeed, is actually being good. No shortcut methods will work (yet—evolution is still going on).

In order to understand why *actually being good* is the most cost-effective solution to this problem, we have to understand it as the price we pay for self-control. I can control myself only with a broad brush. "Moral sentiments may be viewed as a crude attempt to fine-tune the reward mechanism, to make it more sensitive to distant rewards and penalties in selected instances" (Frank 1988, p. 90). As we will see in the next chapter, I cannot micromanage my own real-time deliberations, so I have to resort to shotgun approaches, equipping myself with powerful emotional dispositions that spill over their targets, leave me trembling with rage when rage is appropriate, unable to contain my joy when joy is appropriate, swept away by sorrow or pity. But in order to get these emotions to help me make long-term prudential decisions when I face temptation from short-term Sirens, I have to let them rule me as well when my choice is between my short-term gain and what is best for others. I can't *just* be committed to myself. Or, to put it in terms of my motto, the social environment in which I find myself encourages me, in order to further my *narrow* self-interests, to make myself *larger* than I otherwise would be; when I "look out for Number One" I cast my net wide enough to include my fellow cooperators.

As always, it will not do just to postulate such a happy state of affairs as if it were a gift from God. It might occasionally arise by accident, but if it persists long enough to make a pattern in the world, it needs an explanation. The task of the evolutionary models is to demonstrate that environments can evolve in which this self-enlargement is itself a forced move, rationally dictated. This design "decision"—paying the price of commitment to a variety of *impure* altruism (or is it only advanced benselfishness?) as the cost of gaining self-control—has a rationale that need not have been appreciated by anybody. It is a free-floating rationale, but none the worse for that. In fact, it is *better* as a free-floating rationale. That is what gives emotional expression its evidentiary status in the arms race of detection and dissembling. If we as individuals could readily figure out and act on this rationale, anchoring it in our own minds, we would be suspected of putting on a show. We are highly alert judges of character, and a survey of the cues that matter to us (whether or not we consciously appreciate their contribution) should reveal that we pay scant attention to those displays that are easy to fake, but concentrate instead on the signs that are the irrepressible, unevocable manifestations of disposition. And that is just what we see, Frank claims:

> We can thus imagine a population in which people with consciences fare better than those without. The people who lack them would cheat less often if they could, but they simply have greater difficulty solving the self-control problem. People who have them, by contrast, are able to acquire good reputations and cooperate successfully with others of like disposition. (Frank 1988, pp. 82–83)

And where does this leave the contrast between benselfishness and genuine altruism? Frank claims that the innovation he describes does cross the finish line and get us all the way to genuine altruism:

> People with genuine moral sentiments are better able than others to act in their own interest. . . . People with good reputations can thus solve even nonrepeated prisoner's dilemmas. For example, they can cooperate successfully with one another in ventures where cheating is impossible to detect. Genuine altruism can emerge, in other words, merely on the basis of having established a reputation for behaving in a prudent way. (p. 91)

He shows that in fact altruists—if these good folk really are altruists—do quite well, in spite of the costs they incur. Psychologists and economists have conducted many experiments in which human beings (typically, college students) are put in multiple Prisoner's Dilemmas where the payoffs are small, but not negligible, sums of money. In experiments that Frank conducted, students were given varying opportunities to get to know each other during brief encounters (ten minutes to half an hour) prior to being paired up repeatedly in Prisoner's Dilemma interactions. By varying the conditions, Frank showed that people are surprisingly good—though far from perfect:

between 60 and 75 percent accurate—at predicting who will defect and who will cooperate.

> The prisoner's dilemma experiment lends support to our intuition that we can identify unopportunistic persons. That we can, in fact, do this is the central premise upon which the commitment model is based. From this premise, it logically follows that unopportunistic behavior will emerge and survive even in a ruthlessly competitive material world. We may thus concede that material forces ultimately govern behavior, yet at the same time reject the notion that people are always and everywhere motivated by material self-interest. (Frank 1988, p. 145)

> As the rationalists emphasize, we live in a material world and, in the long run, behaviors most conducive to material success should dominate. Again and again, however, we have seen that the most adaptive behaviors will not spring directly from the quest for material advantage. Because of important commitment and implementation problems, that quest will often prove self-defeating. In order to do well, we must sometimes stop caring about doing the best we can. (p. 211)

Several features of Frank's account support striking correctives to the prevailing philosophical wind encountered in earlier chapters. First, recall the discussion in Chapter 4 of "could have done otherwise," and the example of Martin Luther. Far from such phenomena being exceptions to the rule, or special cases requiring special excuses, we can now see that the practice of making oneself so that one could not have done otherwise is a key innovation in the evolutionary ascent through Design Space—the Vast multidimensional space of all possible designs—to human free will. This tactic of fixing one's will, once recognized, can be seen to have left a fossil trace in one of the words of moral praise that seldom gets brandished by philosophers but is often admired in a moral agent: She shows such *determination*, we say, admiringly. Second, we have seen that the philosophers' fear that if we are determined, we may not be able to avail ourselves of real opportunities—if we are determined there may not *be* any real opportunities—gets it almost backward; we can only be free in a morally relevant sense if, in fact, we learn how to render ourselves *insensitive* to many of the

render selves in sensitive to came
opps that our way

opportunities that come our way. Again, we don't do this by making
ourselves crazy or blind, but by raising our stakes so that the "decisions"
are forced moves, or no-brainers, beneath serious consideration. Third,
we have seen that that mythical being, the economists' purely selfish
rational agent who can never resist a bargain, is a rational fool, to
whom we can put the famous rhetorical question: "If we're so stupid
how come we're rich?" As Frank puts it,

> Altruists . . . do appear to do better economically: the experi-
> mental studies consistently find that altruistic behavior is positively
> correlated with socioeconomic status. Of course, this does not
> mean that altruistic behavior necessarily *causes* economic success.
> But it does suggest that an altruistic posture cannot be too seri-
> ously burdensome in material terms. (Frank 1988, p. 235)

To another mythical being, the Kantian rational saint, we can say in
the same spirit: "If we're so immoral, how come we have so many
trusting friends?" In other words, if you want to get to genuine altruism, you should consider trying the evolutionary approach, sneaking
up on it by gradual increments, with no Prime Mammals, and no skyhooks, passing from blind selfishness through pseudo-altruism to quasi-
altruism (benselfishness) to something that may be quite good enough
for all of us.

    Let me reflect briefly on the methods I have commended on
this path, and the conclusions I am *not* drawing. Frank's arguments and
conclusions have not yet won anything like general acceptance among
his fellow economists or evolutionary theorists (or philosophers), and
there remain serious problems—and alternatives—that need to be carefully addressed. What is mainly important to me here is that Frank's
project, like Ainslie's, is an instance of a *type* of approach to these issues,
a Darwinian approach, that, I claim, is both obligatory and promising.
It is obligatory because any theory of ethics that just helps itself to a
handy set of human virtues without trying to explain how they might
have arisen is in danger of positing a skyhook, a miracle that "explains"
nothing because it can "explain" anything. It is promising because,
contrary to what the enemies of Darwinian approaches declare, novel
insights tumble out of the exercises of these theorists with quite gratifying frequency. Speculative exercises in agent-design have been a sta-
ple of philosophers since Plato's *Republic*. What the evolutionary

perspective adds is a fairly systematic way to keep the exercises naturalistic (so we don't end up designing an angel or a perpetual motion machine), but just as important, it permits us to explore the interactions over time between agents that philosophers typically just handwave about. For instance, philosophers often ask "What if everybody did it?" as a rhetorical question, and don't stop to consider the answer, which they typically think is obvious. They never even address the more interesting question: What if *some* people did it? (What percentage, over what time period, under what conditions?) Computer simulations of evolutionary scenarios add further discipline: a way of discovering hidden assumptions of one's models, and a way of exploring the dynamic effects, by "turning the knobs" to see the effect of different settings of the variables. It is important to recognize that these computer simulations are actually philosophical thought experiments, intuition pumps, not empirical experiments. They systematically explore the implications of sets of assumptions. Philosophers used to have to conduct their thought experiments by hand, one at a time. Now they can conduct thousands of variations in an hour, a good way of checking to make sure that the intuitions they pump are not artifacts of some arbitrary feature of the scenario.

We have arrived at a sketch—a sketch only—of a path from the origin of life to the existence of persons, agents whose freedom is both their greatest strength and their greatest problem. We need to look more closely now at what must be going on inside such a human agent when a free decision is made, before turning to an exploration of the implications for the continuing evolution of human freedom.

Chapter 7

The complexities of social life in a species with language and culture generate a series of evolutionary arms races from which agents emerge who exhibit key components of human morality: an interest in discovering conditions in which cooperation will flourish, sensitivity to punishment and threats, concern for reputation, high-level dispositions of self-manipulation that are designed to improve self-control in the face of temptation, and an ability to make commitments that are appreciable by others. Innovations such as these can thrive under specifiable conditions that co-evolve with them, supplanting the myopic "selfishness" of simpler organisms inhabiting simpler niches.

Chapter 8

The emerging picture of a human agent as a swarm of competing interests shaped by evolutionary forces is hard to reconcile with our traditional sense of ourselves as conscious egos or souls or selves, willing our intentional actions by free decisions that must issue from our private sanctuaries in the mind. This tension is nicely exposed in a controversial—and often misinterpreted— experiment by Benjamin Libet, and can be resolved by looking more closely at how a self emerges from the processes that occur in our brains. Correcting these common misapprehensions about the self and the brain also banishes some dark conclusions about the prospects for free will that have gained credence in some quarters.

Notes on Sources and Further Reading

Among the excellent books on evolutionary approaches to cooperation are Brian Skyrms's *Evolution of the Social Contract* (1996); Robert Wright's *The Moral Animal* (1994) and *Nonzero* (2000); Matt Ridley's *The Origins of Virtue* (1996); Kim Sterelny and Paul E. Griffiths's *Sex and Death: An Introduction to Philosophy of Biology* (1999); and, of course, Elliott Sober and David Sloan Wilson's *Unto Others* (1998). For valuable commentary on Sober and Wilson's book (and a reply), see Katz 2000. I have expressed my views of their book in an essay forthcoming in *Philosophy and Phenomenological Research* (Dennett forthcoming A), which will also contain several other commentaries and a reply by the authors.

On the simple variety of punishment required to enforce cultural norms, see John Haugeland's *Having Thought* (1999), and my review (Dennett 1999A). Paul Bingham (1999) has developed a bold and controversial theory of human evolution based on the premise that the innovation of simple weapons—sticks and stones—so altered the cost-benefit trade-off or riskiness of individual participation in group punishment of defectors that it ushered in the unique varieties of human social cooperativity on which human culture depends, a culturally evolved revolution that was swiftly responded to genetically, with skeletal adaptations for better throwing and weapon-wielding.

The Zahavi Handicap Principle is discussed at length in Frank 1988. See also Helena Cronin's *The Ant and the Peacock* (1991). Randolph Nesse has edited an outstanding anthology of new work on the topic of commitment, *Evolution and the Capacity for Commitment* (2001).

For an overview of the experimental literature on self-manipulation and self-control in children, see J. Metcalfe and W. Mischel's "A Hot/Cool System Analysis of Delay of Gratification: Dynamics of Willpower" (1999). For an overview of the game-theoretic context of Frank's proposal, along with subtle criticism and a friendly amendment to his invocation of emotions for this signaling role, see Don Ross and Paul Dumouchel's "Emotions as Strategic Signals."

Chapter 8

ARE YOU OUT
OF THE LOOP?

*"You imagine a fictional mental construct called 'free will,' which is kind of like
believing in leprechauns or UFOs to a cognitive neuroscientist."*

—Rachel Palmquist, a character in *Brain Storm,* by Richard Dooling

Several years ago I had a strange experience. I was reading a funny and
thought-provoking novel by Richard Dooling, entitled *Brain Storm*
(1998), recommended to me by a friend who insisted I would enjoy it
in spite of its title—in 1978 I had published a book entitled *Brainstorms*.

Drawing the Wrong Moral

The hero of this novel is a young lawyer who visits a neuroscience laboratory in his quest to establish that his client, on trial for murder, has
brain damage. The neuroscientist he finds to help him, Dr. Rachel
Palmquist, is—wouldn't you know—as uninhibited as she is beautiful,
and eventually things get steamy. Their clothing is cast aside, but then,
entwined on the laboratory floor, they encounter a problem: Our hero,
it seems, has a conscience, and thoughts about his wife and kids
threaten to bring the carnal proceedings to an abrupt end. What to do?
Dr. Palmquist does what I guess any brilliant, naked neuroscientist
would do under just such circumstances: she says,

> "In *Consciousness Explained,* Dan Dennett uses the analogy of a
> cartoon featuring Casper the Friendly Ghost. You want to say that
> you have a soul." (Dooling 1998, p. 228)

Free will is the issue, and according to her, I have explained that it cannot exist.

"We don't even have free will?"

"Folk psychology again," she said. "It's a nice fiction. Perhaps a necessary fiction—that a certain part of your consciousness can stand aside from itself, assess and control its own performance. But a brain is a symphony orchestra without a conductor. Right now we're hearing an oboe or maybe a piccolo make an inquisitive flourish of self-examination while the rest of the instruments are off soaring in a different crescendo. What's left of you is an extremely complex balance of competing wet biological parallel processors in that electrochemical batch of elbow macaroni fermenting between your ears, which is ultimately in charge of your body, but by definition cannot be in charge of itself." (Dooling 1998, p. 229)

Quite a wake-up call! This neuroscientist must indeed be brilliant, since she goes on to give an impromptu précis of my theory of consciousness that is insightful and accurate—hard enough to do with clothes on and a podium to stand behind—but what galvanized me was Dooling's master twist: She gets the part about free will dead wrong, *just the way some real neuroscientists have done.* Is free will a fiction, then, according to my view? Is this the implication of my theory of consciousness? Not at all, but more than a few neuroscientists and psychologists have thought that their science has demonstrated this, and my allusion to Casper the Friendly Ghost may have contributed to this misapprehension.

It is easier to see what the issue is if we switch fantasies for a moment. Recall the myth of Cupid, who flutters about on his cherubic wings making people fall in love by shooting them with his little bow and arrow. This is such a lame cartoonists' convention that it's hard to believe that anybody ever took any version of it seriously. But we can pretend: Suppose that once upon a time there were people who believed that an invisible arrow from a flying god was a sort of inoculation that caused people to fall in love. And suppose some killjoy scientist then came along and showed them that this was simply not true: No such flying gods exist. "He's shown that nobody ever falls in love, not *really.* The idea of falling in love is just a nice—maybe even a necessary—fiction. It never happens." That is what some might say. Others, one hopes, would want to deny it: "No. Love is quite real, and *free will morally important*

so is falling in love. It just isn't what people used to think it is. It's just as good—maybe even better. True love doesn't involve any flying gods." The issue of free will is like this. If you are one of those who think that free will is only *really* free will if it springs from an immaterial soul that hovers happily in your brain, shooting arrows of decision into your motor cortex, then, given what *you* mean by free will, my view is that there is no free will at all. If, on the other hand, you think free will might be morally important without being supernatural, then my view is that free will is indeed real, but just not quite what you probably thought it was.

Since readers fall into both camps, you can't hope to reach everybody unless you draw everybody's attention to this problem, which I've often tried to do. In my book *Brainstorms,* one of the questions discussed was whether such things as *beliefs* and *pains* were "real," so I made up a little fable about people who speak a language in which they talk about being beset by "fatigues" where you and I would talk about being tired, exhausted. When we arrive on the scene with our sophisticated science, they ask us which of the little things in the bloodstream are the fatigues. We resist the question, which leads them to ask, in disbelief: "Are you denying that fatigues are *real?*" Given their tradition, this is an awkward question for us to answer, calling for diplomacy (not metaphysics). In *Consciousness Explained* (1991A), I tried to fend off the same confusion with a story about a madman who said there were no animals in the zoo—he knew perfectly well that there are giraffes and elephants and the like, but insisted that they were not what people thought they were. These exercises in imagination-shifting seemed to me to do the trick, but I must say that the message just doesn't seem to take. I've finally come to realize that many people *like* the confusion. They don't want to adjust their imaginations. They like to say that I deny the existence of consciousness, that I deny the existence of free will. Even such a clever thinker as Robert Wright finds the denial of the distinction I insist upon irresistible:

> Of course the problem here is with the claim that consciousness is "identical" to physical brain states. The more Dennett et al. try to explain to me what they mean by this, the more convinced I become that what they really mean is that consciousness doesn't exist. (Wright 2000, p. 398)

And that wily cultural observer Tom Wolfe notes that E. O. Wilson, Richard Dawkins, and I

> present elegant arguments as to why neuroscience should in no way diminish the richness of life, the magic of art, or the righteousness of political causes. . . . Despite their best efforts, however, neuroscience is not rippling out into the public on waves of scholarly reassurance. But rippling out it is, rapidly. The conclusion people out beyond the laboratory walls are drawing is: *The fix is in! We're all hardwired!* That, and: *Don't blame me! I'm wired wrong!* (Wolfe 2000, p. 100)

Exactly the conclusion Rachel Palmquist wanted to draw on the laboratory floor. Later in this chapter, we will confront the problem head-on, in the title of an excellent new book by the psychologist Daniel Wegner: *The Illusion of Conscious Will* (2002). I think Wegner's account of conscious will is the best I have seen. I agree with it in almost every regard. And I've discussed with him the awkwardness—from my point of view—of his title. I see him as the killjoy scientist who shows that Cupid doesn't shoot arrows and then insists on entitling his book *The Illusion of Romantic Love*. But I appreciate that there are people who will insist that Wegner's title is just right: He *is* showing that conscious will is an illusion. Wegner eventually softens the blow by arguing that conscious will may be an illusion, but responsible, moral action is quite real. And that is the bottom line for both of us. We agree that Rachel Palmquist is wrong when she uses a neuroscientific theory of the will to ground her conclusion that our hero's conscience shouldn't trouble him (since he doesn't have free will, not really). Wegner and I agree on the bottom line; what we disagree on is tactics. Wegner thinks it is less misleading, more effective, to say that conscious will is an illusion, but a benign illusion, even, in some regards, a veridical illusion. (Isn't this a contradiction in terms? Not necessarily; like a splittable atom, a veridical illusion can find a place in our conceptual scheme, in spite of its etymology.) I myself think that the temptation to misread this conclusion the way Rachel Palmquist does is so strong that I prefer to make *the same points* by saying that no, free will is *not* an illusion; all the varieties of free will worth wanting are, or can be, ours—but you have to give up a bit of false and outdated ideology to understand how this can be so. Romantic love minus Cupid's arrow is still worth yearning for. It is still, indeed, romantic love, real romantic love.

> **CONRAD:** No, it isn't! Romantic love without genuine spirituality—what you're lampooning as Cupid's arrow— isn't real romantic love at all! It's a cheesy substitute! And the same holds for free will. What *you* call free will, a phenomenon that in the end is just a complicated snarl of mechanistic causes that *look like* decision-making (from certain angles), isn't *real* free will at all!

Fair enough, Conrad, if that's the way you insist on using the terms. But then you must accept the burden of demonstrating why you are wise to hold out for these "genuine" varieties of romantic love and free will, when my substitutes fulfill all the requirements you've listed so far. What makes the "genuine" varieties worth caring about at all? I agree that margarine isn't real butter, no matter how good it tastes, but if you insist on real butter at any price, you really ought to have a good reason.

> **CONRAD:** Aha! You admit it, then. You're just playing with words, and trying to pass off margarine for real butter. I exhort all people to demand real free will; accept no substitutes!

And do you also advise diabetics to insist on "real" insulin, instead of the "artificial" stuff? If your real heart gives out some day, will you spurn an artificial substitute that can perform all the functions of your real heart? At what point does love of tradition turn into a foolish superstition? I claim that the varieties of free will I am defending are worth wanting precisely because they play all the *valuable* roles free will has been traditionally invoked to play. But I cannot deny that the tradition also assigns properties to free will that my varieties lack. So much the worse for tradition, say I.

Perhaps time will tell which expository tactic, Wegner's or mine, is best for the topic of free will, or perhaps not. But shame on anybody who ignores the claim—explicitly defended by both of us—

that a naturalistic account of decision-making still leaves plenty of room for moral responsibility.[1]

What in particular about the neuroscience of decision-making convinces so many people that free will is an illusion? It isn't just the bare fact of materialism—the fact that there are no Cupids shooting arrows into our motor cortex—but rather a particular aspect of that neuroscience, and Rachel Palmquist does a fine job of conveying the popular impression:

> Preconscious cognition is brain activity that occurs *before* you are aware of it. The scary part is that it initiates actual movement in the physical world. Your consciousness, if you want to call it that, simply observes activity which originates somewhere else in your brain. . . . Think of your brain as a complex arrangement of networks and parallel processors. From time to time, some are conscious of themselves, but most aren't. Imagine a three-hundred-millisecond moral void which opens just after the brain triggers behavior and before the brain becomes consciously aware of it. (Dooling 1998, p. 120)

That 300-millisecond "moral void" is the problem. It looks as if your brain makes up its mind before you do!

> "Stimuli, sensations," she said, pasting an electrode on each shoulder. "They get processed preconsciously, important mental decisions and representations are made before the brain is self-consciously aware of them." (p. 122)

The 300-millisecond "gap" is real enough, but there is something fishy about this way of interpreting it—as a "moral void"—and this is the mistake I want to examine. Again. I discussed it in a chapter of *Consciousness Explained,* but that discussion was obscure and difficult, and needs refreshing. This time, perhaps, the moral of the story will come through clearly—instead of coming through backward, the way it was taken by that brilliant, naked neuroscientist, Rachel Palmquist.

1. Disagreeing with us is Derk Pereboom, whose new book, *Living without Free Will* (2001), arrived as I was putting the finishing touches on this book. He defends the view that "given our best scientific theories, factors beyond our control ultimately produce all of our actions, and that we are therefore not morally responsible for them." He did not at all persuade me, but others who find my book unconvincing may find a valuable ally here.

Whenever the Spirit Moves You

Are decisions voluntary? Or are they things that happen to us? From some fleeting vantage points they seem to be the preeminently voluntary moves in our lives, the instants at which we exercise our agency to the fullest. But those same decisions can also be seen to be strangely out of our control. We have to wait to see how we are going to decide something, and when we do decide, our decision bubbles up to consciousness from we know not where. We do not witness it being made; we witness its arrival. This can then lead to the strange idea that Central Headquarters is not where we, as conscious introspectors, are; it is somewhere deeper within us, and inaccessible to us.

—Dennett, *Elbow Room*

It takes time for a brain to do anything, so whenever *you* do something (whenever your body does something) your brain, which controls your body, has to do something else first. Normally, when you are awake and busy, you are doing several things at once—walking and talking, stirring the pot on the stove while trying to recollect which ingredient goes in next, reading the next measure of the piano part while listening to what the cello is playing and moving your own hands into position for the next cascade of chords, or just reaching for your beer while channel surfing. So much is normally going on, overlapped in time, that it would be difficult to sort out all the dependencies, but it is possible to quiet everything down and isolate a "single" act, just in order to study it. Sit very still for a while, trying not to think of anything at all, and then, for no reason at all except that you want to, flick your right wrist once. A single flick, please, whenever, as we say, the spirit moves you. Call that voluntary, intentional act of yours *Flick!* If we monitor your brain with an array of surface electrodes (on the scalp will do fine—we needn't insert them in your brain), we will find that the brain activity leading up to *Flick!* has a definite and repeatable time course, and a shape. It lasts the better part of a second—between 500 and 1,000 milliseconds—ending when your wrist actually moves (which we can detect by having your wrist break a beam of light aimed at a simple photoelectric cell). The motion of the wrist is preceded by less than 50 milliseconds by activity in the motor nerves descending from the motor cortex of your brain to the muscles in your forearm, but it is preceded by as much as 800 milliseconds—almost a second—by a clearly detectable wave of activity in your brain known as the *readiness potential,* or RP (Kornhuber and Deecke 1965). (See Figure 8.1.)

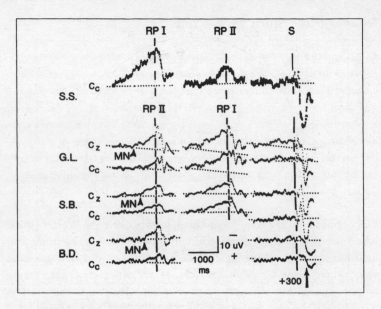

Figure 8.1   EEG Tracings of RP (from Libet 1999, p. 46)

Somewhere among those thousand milliseconds is the notorious "time *t*," the time when *you* consciously decide to flick your wrist. Benjamin Libet set out to determine just when it is. Since this moment is defined by its subjective properties, he had to get *you* to say when it occurs so that he could then superimpose it on the objective series of events occurring in your brain. He figured out a clever way to put the two series, subjective and objective, into registration. He had subjects look at a "clock" with a swiftly moving dot, like the second hand, but moving considerably faster, one revolution every 2.65 seconds, so that he could get readings of fractions of seconds to calibrate against his timed recordings of brain activity (Figure 8.2).

Libet asked his subjects to take note of the position of the dot on the clock face at the instant they decided to flick or were first aware of the urge or wish to flick. This information they were to report (later, well after the flick, without rushing their report). He found a time gap or latency between the RP he measured in subjects' brains and their *reported time of decision* of between 300 and 500 milliseconds. That is the

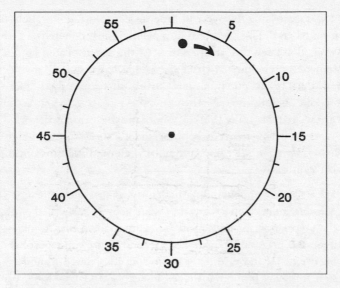

Figure 8.2   Clock Face Used by Libet (from Libet 1999, p. 48)

"moral void" of which Rachel Palmquist speaks, and it is whopping, by neuroscientific standards—compared, for instance, to the idiosyncrasies and inaccuracies that can be observed in other judgments of simultaneity. There is no controversy about whether, in this artificial circumstance, the RP is the triggering cause of your flick. The RP is a highly reliable predictor of flicking. So now what is the problem? It seems to be this: When you *think* you're deciding, you're actually just passively watching a sort of delayed internal videotape (the ominous 300-millisecond delay) of the *real* deciding that happened *unconsciously* in your brain quite a while before "it occurred to you" to flick. As I put it in *Consciousness Explained,*

> We are not quite "out of the loop" (as they say in the White House), but since our access to information is thus delayed, the most we can do is intervene with last-moment "vetoes" or "triggers." Downstream from (unconscious) Command Headquarters, I take no real initiative, am never in on the birth of a project, but do exercise a modicum of executive modulation of the formulated policies streaming through my office. (Dennett 1991A, p. 164)

But I was expressing this view in order to demonstrate its falsehood. I went on to say: "This picture is compelling but incoherent." Others, however, don't see this incoherence. As the sophisticated (and well-dressed) neuroscientist, Michael Gazzaniga has put it: "Libet determined that brain potentials are firing three hundred and fifty milliseconds before you have the conscious intention to act. So before you are aware that you're thinking about moving your arm, your brain is at work preparing to make that movement!" (Gazzaniga 1998, p. 73). William Calvin, another fine (and reliably clothed) neuroscientist puts it more cautiously:

> My fellow neurophysiologist Ben Libet has, to everyone's consternation, shown that the brain activity associated with the preparation for movement (something called the "readiness potential") . . . starts a quarter of a second before you report having decided to move. You just weren't yet conscious of your decision to move, but it was indeed under way. (Calvin 1989, pp. 80–81)

And Libet himself has recently summarized his own interpretation of the phenomenon thus:

> The initiation of the freely voluntary act appears to begin in the brain unconsciously, well before the person consciously knows he wants to act! Is there, then, any role for conscious will in the performance of a voluntary act? (see Libet 1985) To answer this it must be recognized that conscious will (W) does appear about 150 msec. before the muscle is activated, even though it follows the onset of the RP. An interval of 150 msec. would allow enough time in which the conscious function might affect the final outcome of the volitional process. (Actually, only 100 msec. is available for any such effect. The final 50 msec. before the muscle is activated is the time for the primary motor cortex to activate the spinal motor nerve cells. During this time the act goes to completion with no possibility of stopping it by the rest of the cerebral cortex.) (Libet 1999, p. 49)

Only a tenth of a second—100 milliseconds—in which to issue presidential vetoes. As the astute (and impeccably attired) neuroscientist Vilayanur Ramachandran once quipped, "This suggests that our conscious minds may not have free will, but rather 'free won't'!" (Holmes 1998, p. 35). I hate to look a gift horse in the mouth, but I certainly want more free will than that. Can we find any flaws in the reasoning that has led this distinguished group of neuroscientists to this dire conclusion?

Libet's experimental task is an unusual one, worth imagining carefully. You are sitting there calmly, watching a clock dot go round and round, and waiting till, for no reason at all except perhaps that you're getting bored, you decide to flick: "Let the urge to act appear on its own any time without any preplanning or concentration on when to act" (Libet et al. 1983, p. 625). It's important that you *not* follow a policy such as deciding that you'll flick your wrist the next time the clock hand gets to the "three o'clock" position, since then you would have made your decision ("of your own free will") earlier, and just be implementing it more or less mindlessly, triggered by the visual appearance of the clock face. (Recall Martin Luther, who made up his mind long ago, and now can do no other.) How can you be sure you're *not* letting something about the clock face trigger your "free" choice? That is anybody's guess, but for the moment let's presume that you succeed in following instructions at least to this extent: *So far as you can tell,* you are not "gearing" your choice to the position of the clock dot, but rather just "noticing" what position the clock dot is in when "it occurs to you" to flick. After the flick you tell Libet what that position was ("The clock dot was just after 10 when I decided" or "The dot was straight down, position 30" or whatever), and his earlier data-recording permits him to say to the millisecond just when the clock dot was in that position. Libet can then put your stream of consciousness (as later reported by you) into temporal registration with your brain activity, and that will fix the time of your consciousness of your decision, right? That's the assumption that underlies Libet's experiment, but it's not as innocent as it first appears.   *flawed exper.*

Suppose Libet knows that your readiness potential peaked at millisecond 6,810 of the experimental trial and the clock dot was straight down (which is what you reported you saw) at millisecond 7,005. How many milliseconds should he expect to have to add to this number to get the time when you were conscious of it? The light gets from the clock face to your eyeball almost instantaneously, but the path of the signals from retina through lateral geniculate nucleus to striate

cortex takes 5 to 10 milliseconds—a paltry fraction of the 300 millisecond offset, but how much longer does it take for them to get to *you?*
(Or are you located in the striate cortex?) The visual signals have to be
processed before they arrive at wherever they need to arrive at for you
to make a conscious decision of simultaneity. Libet's method presupposes, in short, that we can locate the *intersection* of two trajectories

- the rising-to-consciousness of signals representing the
  decision to flick

- the rising-to-consciousness of signals representing successive clock-face orientations

so that these two events occur side by side, as it were, in a place where
their simultaneity can be noted. Since Libet wants to hear from *you,*
not your striate cortex, we have to know where *you* are in the brain
before we can even begin to interpret the data. Let us suppose, for the
sake of argument, that this makes sense. To be fair and constructive,
cast aside all the extravagant versions of the supposition: Libet is not
supposing that *you* are an actual homunculus, with arms and legs, eyes
and ears, like the little green man in the control room of the man-size
puppet in the morgue in *Men in Black,* and he's not supposing that *you*
are an immaterial portion of glowing ectoplasm that oozes around in
your brain like a ghost amoeba, or that *you* are an angel whose wings
are folded till you are called to fly to heaven. We must consider a min-
imalist version of the hypothesis, stripped of all such embarrassing
details: *You* are just whatever-it-takes-to-be-able-to-experience-
decision-and-clock face-orientation-simultaneity. (If we need to have
an image, we can dimly imagine that this whatever-it-is is some nexus
or cluster of brain activity, and it might shift around under various conditions, a brainstorm with rather special cognitive powers. See Figure
8.3.) There are then at least three possibilities to explore:

(A) You are busy making your free decision in the *faculty of prac-
    tical reasoning* (where all free decisions are made), and you
    have to wait there for visual contents to be sent over from
    the *vision center.* How long does this take? If time pressure
    is not critical, perhaps the visual content is sent very slowly
    and is seriously out of date by the time it arrives, like yes-
    terday's newspaper.

(B) You are busy watching the clock in the *vision center,* and have to wait for the *faculty of practical reasoning* to send you the results of its latest decision-making. How long does this take? This might be another dawdling transmission, mightn't it?

(C) You are sitting where you always sit: in *command head-quarters* (otherwise known as the Cartesian Theater), and have to wait for both the *vision center* and the *faculty of practical reasoning* to send their respective outputs to this place, where everything comes together and consciousness happens. If one of these outposts is farther away, or transmits at a slower rate, you will be subject to illusions of simultaneity—if you judge simultaneity by actual arrival time at command headquarters, instead of relying on something like postmarks or time stamps.

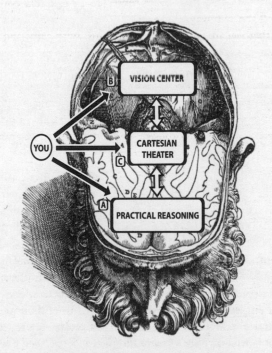

Figure 8.3 Where are You in your brain?

Putting the matter this baldly helps—I hope—to clarify the problems with Libet's picture. What is the presumed implication of these different hypotheses? What would it mean for you to be in one of these places rather than the other? The governing idea is presumably that you can only *act* where you *are*, so if you are not *in* the faculty of practical reasoning when a decision is made there, *you* didn't make it. At best you delegated it. ("I want to be *in* the faculty of practical reasoning. After all, if I'm not *there* when decisions are made, the decisions won't be mine. They will be *its!*") But when you are there, you may get so engrossed in making your decision that "your eyes glaze over" and the vision center's good work goes unattended, never getting to you at all. So, perhaps, you should move back and forth between the faculty of practical reasoning and the vision center. But if that is what you do, then it is quite possible that you were, in fact, conscious of the decision to flick *at the very moment you made it,* but it then took you more than 300 milliseconds to move to the vision center and pick up an image—you got there just as the dot-straight-down picture arrived—so you misjudged the simultaneity because you lost track of how long it took you to get from place to place. Whew! This is one hypothesis, call it *Strolling You,* that could save free will, by showing that the gap was an illusion, after all. According to this hypothesis, you *consciously* decided to flick when that part of your brain decided to flick (hey, you were *there, at the time,* riding the readiness potential as it was created) but you later misjudged the objective clock time of that decision because of the time it took you to get to the vision center and pick up the latest clock-face position.

If you don't like that hypothesis, here is another one that could do the trick, based on alternative (C), in which both the vision center and the faculty of practical reasoning are moved out of command headquarters. Call it *Out-of-touch You.* You have outsourced all these tasks, as today's business world would put it, delegating them to subcontractors, but you do keep limited control of their activities from your seat in command headquarters by sending them orders and getting results from them, in a continuous cycle of commands and responses. If asked to think of a reason not to dine out tonight, you send out to your faculty of practical reasoning for a reason, and pretty quick it sends two back: *I'm too tired* and *there's food in the fridge that will spoil if we don't eat it tonight.* How did the faculty come up with these? Why in this order?

What operations did it execute to generate them? You haven't a clue—you just know what you sent out for, and recognize that what arrived back is a satisfactory fulfillment of your request. If asked what time it is, you send the appropriate command to the vision center, and it sends back the latest view of the watch on your wrist, with a little help from the *wrist-motion-control center,* but you have no insight into how that collaborative effort was achieved either. Given the problem of variable time delays, you institute a time-stamp system, which works well for most purposes, but you misuse it in Libet's rather unnatural setting. When asked, from your *underprivileged* position in command headquarters to judge just when, exactly, your faculty of practical reasoning issued its flick order (a judgment you are to render in terms of the time stamps you discern on the streams of reports coming in from both the faculty of practical reasoning and the vision center), you match up the wrong reports. Since you're relying on second-hand information (reports from the two outlying subcontractors) you can easily just be wrong about which event happened first, or whether any two were simultaneous.

One thing going for this hypothesis is that such judgments of simultaneity are unnatural acts in the first place, unless they are framed for a particular purpose, such as your trying to get your staccato attack in sync with the conductor's downbeat, or trying to connect with a low fastball so as to send it straight back over the pitcher's head. In such natural contexts, virtuoso feats of timing are possible, but isolated judgments of "cross-modal" simultaneity (answering such questions as "Which came first, the flash or the beep, or were they simultaneous?") are notoriously prone to interference and error. Depending on how you frame a judgment, depending on what use you plan to make of the judgment, what counts, subjectively, as simultaneity can be made to move around. So if you make your judgments of simultaneity from such an underprivileged position, with no natural context that provides a reason for the judgment, you could well have ordered the faculty of practical reason to issue a decision, and simply misfiled its report of completion, so that you misjudge it to have been done simultaneously with the perception in the vision center of the clock face position at 30. But perhaps this hypothesis doesn't appeal, since *you* aren't actually *present* in the faculty of practical reasoning when *it* makes the decision.

So here is yet another hypothesis, which puts you back where the action is (or was): _Slow-drying Ink_. When you make a decision, consciously, _in_ the faculty of practical reasoning (and you are right there, in the thick of it), you "write it up" in ink that is slow-drying: Although you can start acting on it immediately, you can't compare it with what's going on in vision until the ink dries (in about 300 milliseconds). (This hypothesis is inspired by other work of Libet's discussed in _Consciousness Explained_ [Dennett 1991A], on "backwards referral" of consciousness.) On this hypothesis, _you_ actually decide to execute _Flick!_ exactly when the RP in your brain shows up, without any delay, but you don't get to compare that conscious decision with a result from the vision center for a good 300-plus milliseconds, the time it takes for your decision to cure before entering the comparison chamber.

And if you don't like that hypothesis, there are others that could be considered, including, of course, all manner of hypotheses that _don't_ "save free will" because they tend to confirm Libet's view of the matter: that in the normal course of moral decision-making, _you_ in fact have at most 100 milliseconds in which to veto or otherwise adjust decisions made earlier (and elsewhere) unconsciously. Can't we just dismiss the whole sorry lot of them, on the grounds that these hypotheses are wildly unrealistic oversimplifications of what is known about how decision-making works in the brain? Yes indeed, we could, and we should. But when we do that, we don't just dismiss all these fanciful hypotheses that could "save free will" in the face of Libet's data; we must also dismiss Libet's own hypothesis and all the others that purport to show we only have "free won't." His hypothesis, just as much as those I've just sketched, depends on taking seriously the idea that _you_ are restricted to the materials you can get access to from a particular subregion of the brain. How so? Consider his idea of a strictly limited window of opportunity to veto. Libet tacitly presupposes that _you_ can't start thinking seriously about whether to veto something until you're conscious of what it is that you might want to veto, and you have to wait 300 milliseconds or more for this, which gives you only 100 milliseconds in which to "act": "This provides a period during which the conscious function could potentially determine whether the volitional process will go on to completion" (Libet 1993, p. 134). The "conscious function" waits, in the Cartesian Theater, until the information arrives, and only then *for the first time* has access to it and can start thinking about what to do about it, whether to veto it, etc. But why couldn't *you* have been thinking ("unconsciously") about whether to veto *Flick!* ever since *you* decided ("unconsciously") to flick, half a second ago? Libet must be assuming that the brain is talented enough to work out the details of implementation on how to flick over that period of time, but only a "conscious function" is talented enough to work on the pros and cons of a veto decision.

In fact, at one point Libet sees this problem and addresses it candidly: "The possibility is not excluded that factors, on which the decision to veto (control) is *based,* do develop by unconscious processes that precede the veto" (Libet 1999, p. 51). But if that possibility is not excluded, then the conclusion Libet and others should draw is that the 300-millisecond "gap" has *not* been demonstrated at all. After all, we know that in normal circumstances the brain begins its discriminative and evaluative work as soon as stimuli are received, and works on many concurrent projects at once, enabling us to respond intelligently just in time for many deadlines, without having to stack them up in a queue waiting to get through the turnstile of consciousness before evaluation begins. Patricia Churchland (1981) demonstrated this in a simple experiment in which subjects were required to respond consciously (how else?) to a light flash. Their *total* response time was about 350 milliseconds. Libet's reaction to Churchland's finding was to insist that such a response is begun unconsciously: "The ability to detect a stimulus and react to it purposefully, or to be psychologically influenced by it, without any reportable conscious awareness of the stimulus, is widely accepted" (Libet 1981, p. 188). But this concedes just what is at issue: *You* can begin reacting purposefully to—*you* can be psychologically influenced by—a decision to flick long before it "rises to consciousness." For all Libet's experiments have shown, it could be that you have *optimal* access at all times to the decision-making you are engaged in. That is, it could be that every part of you that is competent to play any role in the decision-making it falls to you to engage in gets whatever it needs to do its job at the earliest possible time. (What else could you be worried about when you wonder if *you* are getting informed too late to make the difference you want to make?)

Libet's data do rule out one hypothesis, which might have been our favorite: *Self-contained You,* according to which *all* the brain's chores are gathered into one compact location, where everything could happen at once in one place—vision, hearing, decision-making, simultaneity-judging. . . . With everything so handy, the timing problem couldn't arise: A person, a soul, could sit there and make free, responsible decisions and be simultaneously conscious of making them, and of everything else going on in consciousness at the time. But there is no such place in the brain. As I never tire of pointing out, all the work done by the imagined homunculus in the Cartesian Theater has to be broken up and distributed in space and time in the brain. It is once again time to repeat my ironic motto: If you make yourself really small, you can externalize virtually anything.

The brain processes stimuli over time, and the amount of time depends on which information is being extracted for which purposes. A top tennis player can set up to design a return of service within 100 milliseconds or so. The 78 feet from base line to base line can be traversed by a serve from Venus Williams (averaging 125 mph) in less than 450 milliseconds, only about 50 milliseconds more than it took the fastest serve yet recorded (from Greg Rusedski, at 147 mph initial speed). And since the precise timing and shape of that return depends critically on visual information (if you doubt this, try returning service blindfolded), it is possible for the brain to extract visual information and put it to highly appropriate use in that short a time. As Churchland showed, just pressing a button when asked to signal when you see a flash of light takes a normal subject about 350 milliseconds. Now these are conscious, voluntary, intentional responses to events (aren't they?), and they happen without any 300 to 500 millisecond delay. Of course, the tennis player, and the subject in the experiment, have to decide (freely, consciously) beforehand that they are going to gear their responses to particular conditions. These are, in effect, mini-Luther cases. The tennis player pre-commits to a simple plan and then lets "reflexes" execute her intentional act. (It can be *somewhat* conditional, along the lines of *IF high to my backhand THEN defensive lob ELSE topspin down alley*. In effect, she turns herself temporarily into a *situation-action machine*.) And you, having decided to cooperate with the experimenter by pressing the button as soon as the light flash appears, do likewise: You just sit back on autopilot and let your decision be implemented. "I couldn't do otherwise," you might say. "Since there wasn't time to reflect and consider, I did all my reflection off-line, in the luxury of spare time, so that when the crunch came I could act without thinking."

    We do this all the time. Our lives are full of decisions to act when the time is ripe, revisable commitments to policies and attitudes that will shape responses that must be delivered too swiftly to be reflectively considered in the heat of action. We are the authors and executors of these policies, even though they are compiled from parts we can only indirectly monitor and control. The fact that we can play ensemble music, for instance, shows that our brains are capable of multitasking on a highly convoluted timescale, and it is all deliberate, controlled, and intended. The responses we make in conversation, indeed the very words we say silently *to ourselves* as we reflect on what to do next, are themselves acts that have had long preparation times reaching back into the past. What Libet discovered was not that consciousness lags ominously behind unconscious decision, but that conscious decision-making takes time. If you have to make a series of conscious decisions, you'd better budget half a second, roughly, for each one, and if you need to control things faster than that, you'll have to compile your decision-making into a device that can leave out much of the processing that goes into a stand-alone conscious decision. Libet reports a simple experiment by Jensen (1979) that demonstrates this effect. Jensen asked subjects to press a button as soon as they were conscious of a light flash, just as Patricia Churchland had done, and got results consonant with hers—actually, his subjects' reaction times were quite a bit faster—250 milliseconds on average. Then he asked his subjects to delay their button presses just a little, *as little as possible*. They had to add a whopping 300 milliseconds to their response time. The brain has tricks for avoiding these delays under some conditions, such as searching a scene for particular items under time pressure. For instance, when hunting for a target item, the brain sometimes knows enough to let itself go; it does a random visual search of a systematic display, even though it could do a "more efficient" methodical search. Attention can swing faster from item to item when it is just let loose, since "attention is fast, but volition is slow" (Wolfe, Alvarez, and Horowitz, 2000).

    These timing tricks usually fit together seamlessly and are incorporated into the brain's own monitoring of what it is up to, but in artificial circumstances (as devised by clever experimenters) the tricks can be exposed. For instance, when the brain executes a decision to act (at the time of the RP rise) it sets up anticipations—it produces a little future—about what should happen next. If what happens next is artificially disrupted—by being sped up or delayed, for instance—this creates violations of those anticipations and signals that something is wrong. But the brain may not be up to coming up with the right interpretation of just what has happened in such an unprecedented setting. In *Consciousness Explained* (Dennett 1991A, pp. 167–68), I described an early experiment that illustrates this, which I called Grey Walter's pre-cognitive carousel. Back in the early 1960s, the eminent neurosurgeon and early roboticist Grey Walter took advantage of the fact that he had a series of epilepsy patients in whose motor areas he had implanted electrodes. He wired the leads from the electrodes to a slide carousel, so that whenever the patients decided (*ad lib,* whenever the spirit moved them) to advance to the next slide, the detected brain activity in the motor area directly triggered the advance of the carousel. The button the patients pushed was a dummy, attached to nothing. The effect, he said, was dramatic: It seemed to the patients that just as they were "about to" push the button, but *before* they had decided, the slide projector would read their minds and literally take matters out of their hands.[2] Since their anticipation of a perceived slide change was "scooped" by slightly earlier perception of such a change, they were left with a powerful conviction that something spooky was happening; the slide projector was reading their minds. In one sense that is just

---

2. Grey Walter described this experiment in a talk I attended in Oxford in 1963 or 1964. The account was never subsequently published, to the best of my knowledge. I and a number of readers have tried to track it down, without success, and several—Wegner included—have expressed the hunch that Grey Walter was pulling our legs that day in Oxford. Maybe, but my own surmise is that he may have decided not to publish it because even by the standards of the day, the ethics of the experiments were borderline: His patients had chronically implanted phone jacks protruding from their skulls for months on end, a regimen they would not likely have acquiesced in, had they not thought it was part of a treatment that might improve their epilepsy, but as best I recall, their repeat visits to Grey Walter's Burden Institute were as research subjects in experiments that had no plausible therapeutic benefit to them. (In any event, the effect should be possible to replicate non-invasively on normal subjects today, with the aid of the latest high-speed analysis of scalp electrode signals or MEG scanning. The main technical hurdle is not getting the data, but processing it fast enough in real time to provide the anticipation effect. Although I know of no published replications—or failures to replicate—I predict that anybody who takes the trouble to test this and the variations I propose on p. 168 of *Consciousness Explained* will find the effect.)

what was happening, but it wasn't learning of their decisions before they were conscious of them—it was just "reading" and executing their conscious decisions faster than their own arm muscles could "read" and execute the very same decisions. Imagine popping a photograph into an envelope and mailing it (snail mail) to a friend, and suppose the letter is swiftly intercepted by a mail thief who, as a prank, scans your photograph and e-mails the picture to your friend minutes after you drop the envelope in the mailbox. Half an hour after you mail the photograph, your friend calls you and marvels at the details of the picture. You were anticipating just such a call, but not for two or three days! It would be upsetting, to say the least, and you might be tempted to jump to the false conclusion that your letter must have been sent by you long before you were conscious of sending it—have you been sleepwalking in recent days?

A similar confusion, I submit, is what is happening in the case of Libet's subjects' 300-millisecond misjudgment. When we perform an intentional action, we normally monitor it visually (and by hearing and touch, of course) to make sure it is coming off as intended. Hand–eye coordination is accomplished by a tightly interwoven system of sensory and motor systems. Suppose I am intentionally typing the words "flick the wrist" and wish to monitor my output for typographical errors. Since the motor commands take some time to execute, my brain should *not* compare the *current* motor command with the *current* visual feedback, since by the time I see the word "flick" on the screen, my brain is already sending the command *type "wrist"* to my muscles. My brain should keep the earlier command (*type "flick"*) around long enough (slow-drying ink?) to use it efficiently for visual monitoring purposes. If that habit is sufficiently ingrained (and why wouldn't it be?), it should interfere with the attempt to perform the unnatural act of timing the decision itself rather than the executed action. The only way to get Libet's data to imply an ominous 300-millisecond gap is to assume that the simultaneity judgment he calls for is undistorted by any such habit, but we have good reasons for believing otherwise, so the gap is an artifact of mis-imagined theory, not a discovery.

When we remove the Cartesian bottleneck, and with it the commitment to the ideal of the mythic time *t,* the instant when the conscious decision happens, Libet's discovery of a 100-millisecond veto

[handwritten: don't measure free will in instants]

window evaporates. Then we can see that our free will, like all our other mental powers, has to be smeared out over time, not measured at instants. Once you distribute the work done by the homunculus (in this case, decision-making, clock-watching, and decision-simultaneity-judging) in both space and time in the brain, you have to distribute the moral agency around as well. You are not out of the loop; you *are* the loop. You are that large. You are not an extensionless point. What you do and what you are *incorporates* all these things that happen and is not something separate from them. Once you can see yourself from that perspective, you can dismiss the heretofore compelling concept of a mental activity that is *unconsciously* *begun* and then only later "enters consciousness" (where *you* are eagerly waiting to get access to it). This is an illusion since many of the reactions *you* have to that mental activity are initiated at the earlier time—your "hands" reach that far, in time and space.[3]

[handwritten: you are the loop]

A Mind-writer's View

Illusory or not, conscious will is the person's guide to his or her own moral responsibility for action.

—Daniel Wegner, *The Illusion of Conscious Will*

If Libet's Cartesian Theater sketch model of conscious decision-making is too simple, what does a better model look like? Daniel Wegner's model has the curious virtue of being halfway in the right direction. By being still too Cartesian, still too dependent on the seductive metaphor of "the place in the brain where I am," it illustrates the powerful attractions of that idea. It is very difficult, in fact, to describe the immediate phenomenology of decision-making in any other terms,

---

3. One commentator on Libet who gets close is Sean Gallagher: "I think that this problem can be solved as long as we do not think of free will as a momentary act. Once we understand that deliberation and decision are processes that are spread out over time, even, in some cases, very short amounts of time, then there is plenty of room for conscious components that are more than accessories after the fact" (Gallagher 1998). (But then he goes on to say that if the feedback is all unconscious, it will be "deterministic" but if it is conscious, it won't be. Cartesian thinking dies hard.)

so by taking our bearings at Wegner's halfway house, we can see better how to complete the departure from the Cartesian Theater.

Everybody knows what a mind-reader is supposed to be able to do; Wegner is an accomplished mind-*writer*. He has figured out how to compose intentional actions and impose them on people so that they think they are deciding on their own to do them. There is a cottage industry in the philosophical world of free will research devoted to the analysis of thought experiments involving various imaginary mind-writers, such as nefarious neurosurgeons who implant remote-control devices in their victims' brains, but the truth about actual mind-writing has some wrinkles in it that are, in my view, of greater philosophical interest.

How can anybody write intentions in somebody else's mind? Don't we each have "privileged access" to our own decisions and choices? No, not really. One of the major themes in Wegner's work is the demonstration, by a number of routes, that our knowledge of the relation between our thoughts and our actions (and between thoughts and other thoughts) has only the "privilege" of ordinary familiarity. If I know better than you know what I am up to, it is only because I spend more time with myself than you do. But if you surreptitiously insert grounds for false belief into my stream of consciousness, you can make me think I am making "free" decisions when it is you who controls my actions. The basic technique has been understood by magicians for centuries: Magicians now call it *psychological forcing,* and it is remarkably effective in able hands. You give the victim a variety of reasons to think that he and he alone is responsible for deciding something that you want him to decide, and he falls for it. Or you can fool him in the other direction, getting him to think he is not responsible for something that he is, in fact, producing—for instance, a message spelled out by "spirits" on a Ouija board.

Wegner has adapted the principle of the Ouija board and the techniques of the magicians to a laboratory setting, and produced some remarkable results. Subjects in his experiments are systematically driven to misattribute decisions to themselves that are in fact being made by somebody else. The reason they can be fooled is that, as David Hume pointed out so vigorously several centuries ago, you can't *perceive* causation. You can't see it when it happens outside, and you can't introspect it when it happens inside. What people perceive is one thing happening and then another, and they fall for Wegner's magic for much the same reasons we all fall for stage magic: We are overeager to interpret, to "notice" things causing other things when, in fact, both "cause" and "effect" are effects of complex machinery that is hidden from us—backstage, in effect. He shows that we don't have anything like direct access to the causes and effects of our decisions and intentions, but rather must draw inferences—swiftly and without logical fanfare. We're actually very good at this; the inferences we make are almost always inferences to the best explanation of the sequence we've experienced, except when a sly manipulator has put some misleading premises into the arena.

Notice how the introduction of the issue of privileged access automatically puts us on the slippery slope to the Cartesian Theater: There are things going on in me that I don't know about, and then there are things I know about "directly"—they are somehow delivered to *me* wherever I am. Instead of fighting this, Wegner permits himself the full Cartesian image when it suits his purposes: "We can't possibly know (let alone keep track of) the tremendous number of mechanical influences on our behavior because we inhabit an extraordinarily complicated machine" (Wegner 2002, p. 27). These machines we inhabit simplify things for our benefit: "The experience of will, then, is the way our minds portray their operations to us, then, not their actual operation" (p. 96). In other words, we get a useful but distorted glimpse of what is going on in our brains:

> The unique human convenience of conscious thoughts that preview our actions gives us the privilege of feeling we willfully cause what we do. In fact, unconscious and inscrutable mechanisms create both conscious thought about action and the action, and also produce the sense of will we experience by perceiving the thought as cause of the action. So, while our thoughts may have deep, important, and unconscious causal connections to our actions, the experience of conscious will arises from a process that interprets these connections, not from the connections themselves. (Wegner 2002, p. 98)

Who or what is this "we" that inhabits the brain? It is a commentator and interpreter with limited access to the actual machinery, more along the lines of a press secretary than a president or boss. And this imagery leads straight to Libet's vision of "conscious will" as being out of the loop.

> Consciousness and action seem to play a cat-and-mouse game over time. Although we may be conscious of whole vistas of action before the doings get underway, it is as though the conscious mind then slips out of touch. A microanalysis of the time interval before and after action indicates that consciousness pops in and out of the picture and *doesn't really do anything* [my italics— DCD]. The Libet research, for one, suggests that when it comes down to the actual instant of a spontaneous action, the experience of consciously willing the action occurs only after the RP signals that brain events have already begun creating the action (and probably the intention and the experience of conscious will as well). (Wegner 2002, p. 59)

A Self of One's Own

Odd, though, all these dealings of mine with myself. First I've agreed a principle with myself, now I'm making out a case to myself, and debating my own feelings and intentions with myself. Who is this self, this phantom internal partner, with whom I'm entering into all these arrangements? (I ask myself.)

—Michael Frayn, *Headlong*

Philosophers and psychologists are used to speaking about an organ of unification called the "self" that can variously "be" autonomous, divided, individuated, fragile, well-bounded, and so on, but this organ doesn't have to exist as such.

—George Ainslie, *Breakdown of Will*

A voluntary action is something a person can do when asked.

—Daniel Wegner, *The Illusion of Conscious Will*

So "consciousness . . . doesn't really do anything" according to Wegner, and that is why conscious will is, as his title proclaims, an illusion. There is an escape from this vision, thanks to a slight shift in perspective that is actually implicit in Wegner's work. Consciousness has lots of work to do, but its accomplishments seem to disappear when we ask ourselves what work it is doing *right now* (at time *t*). Since at each moment it "doesn't really do anything," it can seem that it is an utterly epiphenomenal accompaniment, along for a free ride. An evolutionary perspective shows us why this is mistaken.

One of the phenomena that Wegner exposes for a better view is "ideomotor automaticity." This is the name for the familiar—but always unsettling—phenomenon in which thinking about something can bring about a bodily action related to that thing without the action being an intentional action. For instance, you might betray a secret sexual thought with a telltale hand motion that you didn't intend and, in fact, would be embarrassed to discover. In such a case, you are not conscious of the causal relation between the thought and the act, but there it is, as good as the causal relation between the aroma of good food and salivation. The main feature of ideomotor actions is people's obliviousness to them—their *underprivileged* access, you might say. It is as if our usually transparent minds had curtains or barriers installed, behind which these causal chains could get tugged without our introspecting them, producing effects without our compliance. "This ghost army of unconscious actions provides a serious challenge to the notion of an ideal human agent. The greatest contradictions to our ideal of conscious agency occur when we find ourselves behaving with no conscious thought of what we are doing" (Wegner 2002, p. 157).

For Descartes, the mind was perfectly transparent to itself, with nothing happening out of view, and it has taken more than a century of psychological theorizing and experimentation to erode this ideal of perfect introspectability, which we can now see gets the situation almost backward. Consciousness of the springs of action is the exception, not the rule, and it requires some rather remarkable circumstances to have evolved at all. Ideomotor actions are the fossils, in effect, of an earlier age, when our ancestors were not as clued in as we are about what they were doing. As Wegner says, "Rather than needing a special theory to explain ideomotor action, we may only need to explain why ideomotor actions and automatism have eluded the mechanism that produces the experience of will" (p. 150).

In most of the species that have ever lived, "mental" causation has no need for, and hence does not evolve, any elaborate capacity for self-monitoring. In general, causes work just fine in the dark, without needing to be observed by anybody, and that is as true of causes in animals' brains as anywhere else. So however "cognitive" an animal's faculties of discrimination might be, the capacity of their outputs to cause the selection of appropriate behavior does not need to be experienced *by anything or anybody.* A bundle of situation-action links of indefinite sophistication can reside in the nervous system of a simple creature and serve its many needs without any further supervision. Its individual actions may need to be guided by a certain amount of internal self-monitoring (specific to the action), in order to make sure, for example, that each predatory swipe snags its target, or to get the berries into the mouth, or to guide the delicate docking with the sexual parts of a conspecific of the opposite sex, but these feedback loops can be as isolated, as local, as the controls that spur the immune system into action when infection looms, or adjust the heart rate and breathing during exercise. (This is the truth behind the deeply misleading intuition that invertebrates, if not "higher, warm-blooded" animals, might be "robots" or "zombies," altogether lacking minds.)

As creatures acquire more and more such behavioral options, however, their worlds become cluttered, and the virtue of tidiness can come to be "appreciated" by natural selection. Many creatures have evolved simple instinctual behaviors for what might be called home improvement, preparing paths, lookouts, hideouts, and other features of their neighborhoods, generally making the local environment easier to get around in, easier to understand. Similarly, when the need arises, creatures evolve instincts for sprucing up their most intimate environments: their own brains, creating paths and landmarks for later use. The goal unconsciously followed in these preparations is for the creature to come to know its way around itself, and how much of this internal home improvement is accomplished by individual self-manipulation and how much is incorporated genetically is an open empirical question. Along one of these paths, or many of them, lie the innovations that lead to creatures capable of considering different courses of action in advance of committing to any one of them, and weighing them on the basis of some projection of the probable outcome of each. In Chapter 5, we considered the advent of choice machines that were capable of evaluating the probable outcomes of candidate options prior to decision. In the quest by brains to produce useful future, this is a major improvement over the risky business of *blind* trial and error, since, as Karl Popper once put it, it permits some of your hypotheses to die in your stead. Such Popperian creatures, as I have called them, get to test some of their hunches in informed simulations, rather than risking them in the real world, but they needn't understand the rationale of this improvement in order to reap the benefits. The appreciation of the likely effects of particular actions is built into any such assessment, but the appreciation of the effects of the contemplation itself is a still higher, even more optional, level of self-monitoring. You don't have to know you're a Popperian creature to be one. After all, any chess-playing computer considers and discards thousands or millions of possible moves on the basis of their probable outcomes, and it is manifestly not a conscious or self-conscious agent. (Not yet—the future may hold conscious and even self-conscious robots, which are certainly not impossible.)

What was it that arose in the world to encourage the evolution of a *less unwitting* implementation of Popperian behavioral control? What new environmental complexity favored the innovations in control structure that made this possible? In a word, communication. It is only once a creature begins to develop the activity of communication, and in particular the communication of its actions and plans, that it has to have some capacity for monitoring not just the results of its actions, but of its prior evaluations and formation of intentions as well (McFarland 1989). At that point, it needs a level of self-monitoring that keeps track of which situation-action schemes are in the queue for execution, or in current competition for execution—and which candidates are under consideration in the faculty of practical reasoning, if that is not too grand a term for the arena in which the competition ensues. How could this new talent arise? We can tell a Just So Story that highlights the key features.

Compare the situation confronting our ancestors (and Mother Nature) to the situation confronting the software engineers who wanted to make computers more user-friendly. Computers are fiendishly complex machines, most of the details of which are nauseatingly convoluted and, for most purposes, beneath notice. Computer-users don't need information on the states of all the flip-flops, the actual location of their data on the disk, and so forth, so software designers created a series of simplifications—even benign distortions in many cases—of the messy truth, cunningly crafted to mesh with, and enhance, the users' preexisting powers of perception and action. Click and drag, sound effects, and icons on desktops are the most obvious and famous of these, but anybody who cares to dig deeper will find a bounty of further metaphors that help make sense of what is going on inside, but always paying the cost of simplification. As people interacted more and more with computers, they devised a host of new tricks, projects, goals, ways of using and abusing the competences designed for them by the engineers, who thereupon went back to the drawing board to devise further refinements and improvements, which were then used and abused in turn, a coevolutionary process that continues apace today. The user interface we interact with today was unimagined when computers first appeared, and it is the tip of an iceberg in several senses: Not only are the details of what goes on inside your computer hidden, but so are the details of the history of R&D, the false starts, the bad ideas that fizzled before ever reaching the public (as well as the notorious ones that did and failed to catch on). A similar process of R&D created the user interface between talking people and other talking people, and it uncovered similar design principles and (free-floating) rationales. It too was coevolutionary, with people's behaviors, attitudes, and purposes evolving in response to the new powers they discovered. Now people could *do things with words* that they could never do before, and the beauty of the whole development was that it *tended* to make those features of their complicated neighbors that they were most interested in adjusting readily accessible to adjustment from outside—even by somebody who knew nothing about the internal control system, the brain. These ancestors of ours discovered whole generative classes of behaviors for adjusting the behavior of others, and for monitoring and modulating (and, if need be, resisting) the reciprocal adjustment of their own behavioral controls by those others.

The centerpiece metaphor of this coevolved human user-illusion is the Self, which appears to reside in a place in the brain, the Cartesian Theater, providing a limited, metaphorical outlook on what's going on in our brains. It provides this outlook to others, *and to ourselves*. In fact, we wouldn't exist—as Selves "inhabiting complicated machinery," as Wegner vividly puts it—if it weren't for the evolution of social interactions requiring each human animal to create within itself a subsystem designed for interacting with others. Once created, it could also interact with itself at different times. Until we human beings came along, no agent on the planet enjoyed the curious *non*-obliviousness we have to the causal links that emerged as salient once we human beings began to talk about what we were up to.[4] As Wegner puts it, "People become what they think they are, or what they find that others think they are, in a process of negotiation that snowballs constantly" (Wegner 2002, p. 314).

When psychologists and neuroscientists devise a new experimental setup or paradigm in which to test non-human subjects such as rats or cats or monkeys or dolphins, they often have to devote dozens or even hundreds of hours to training each subject on the new tasks. A monkey, for instance, can be trained to look to the left if it sees a grating moving up and look to the right if it sees a grating moving down. A dolphin can be trained to retrieve an object that looks like (or *sounds,* to its echolocating system, like) an object displayed to it by a trainer. All this training takes time and patience, on the part of both trainer and subject. Human subjects in such experiments, however, can usually just be told what is desired of them. After a brief question-and-answer session and a few minutes of practice, we human subjects will typically be as competent in the new environment as any agent ever could be. Of course, we do have to *understand* the representations presented to us in these briefings, and what is asked of us has to be composed of action-parts that fall within the range of things we can do. That is what Wegner means when he identifies voluntary actions as things we can do when asked. If asked to lower your blood pressure or adjust your heartbeat or wiggle your ears, you will not be so ready to comply, though with training not unlike that given to laboratory animals, you may eventually be able to add such feats to your repertoire of voluntary actions.

As Ray Jackendoff has pointed out to me, when language came into existence, it brought into existence the kind of mind that can transform itself on a moment's notice into a somewhat different virtual machine, taking on new projects, following new rules, adopting new policies. We are transformers. That's what a mind is, as contrasted

---

4. Philosophers may want to compare my Just So Story to Wilfrid Sellars's (1963) myth of "our Rylean ancestors," and "Jones, the inventor of thoughts." My debt to Sellars should be clear to them.

with a mere brain: the control system of a chameleonic transformer, a
virtual machine for making more virtual machines. Non-human animals can engage in voluntary action of sorts. The bird that flies wherever it wants is voluntarily wheeling this way and that, voluntarily
moving its wings, and it does this without benefit of language. The distinction embodied in anatomy between what it can do voluntarily (by
moving its striated muscles) and what happens autonomically, moved
by smooth muscle and controlled by the autonomic nervous system, is
not at issue. We have added a layer on top of the bird's (and the ape's
and the dolphin's) capacity to decide what to do next. It is not an
anatomical layer in the brain, but a functional layer, a virtual layer
composed somehow in the micro-details of the brain's anatomy: We
can ask each other to do things, and we can ask ourselves to do things.
And at least sometimes we readily comply with these requests. Yes,
your dog can be "asked" to do a variety of voluntary things, but it can't
ask why you make these requests. A male baboon can "ask" a nearby
female for some grooming, but neither of them can discuss the likely
outcome of compliance with this request, which might have serious
consequences for both of them, especially if the male is not the alpha
male of the troop. We human beings not only can do things when
requested to do them; we can answer inquiries about what we are doing
and why. We can engage in the practice of asking, and giving, reasons.

It is this kind of asking, which we can also direct to ourselves,
that creates the special category of voluntary actions that sets us apart.
Other, simpler intentional systems act in ways that are crisply predictable on the basis of beliefs and desires we attribute to them on the
basis of our surveys of their needs and their history, their perceptual and
behavioral talents, but some of our actions are, as Robert Kane insisted,
self-forming in a morally relevant way: They result from decisions we
make in the course of trying to make sense of ourselves and our own
lives (Coleman 2001). Once we begin talking about what we're doing,
we need to keep track of what we're doing so we can have ready
answers to these inquiries. Language requires us to keep track, but also
helps us keep track, by helping us categorize and (over)simplify our
agendas. We cannot help but become amateur auto-psychologists.
Nicholas Humphrey and others have spoken of apes and other highly
social species as *natural psychologists,* because of the manifest skill and
attention they devote to interpreting each other's behavior, but since,

unlike academic psychologists—and other human beings—apes never get to compare notes, to argue about attributions of motives and beliefs, their competence as psychologists never obliges them to use explicit representations. With us, it is different. We need to have something to say when asked what the heck we think we're doing. And when we answer, our authority is problematic. The evolutionary biologist William Hamilton, reflecting on his own uneasiness with his recognition of this fact, put the issue particularly well:

> In life, what was it I really wanted? My own conscious and seemingly indivisible self was turning out far from what I had imagined and I need not be so ashamed of my self-pity! I was an ambassador ordered abroad by some fragile coalition, a bearer of conflicting orders, from the uneasy masters of a divided empire. . . . As I write these words, even so as to be able to write them, I am pretended to a unity that, deep inside myself, I now know does not exist. (Hamilton 1996, p. 134)

Wegner is right, then, to identify the Self that emerges in his and Libet's experiments as a sort of public-relations agent, a spokesperson instead of a boss, but these are extreme cases set up to isolate factors that are normally integrated, and we need not identify *ourselves* so closely with such a temporarily isolated self. (If you make yourself really small, . . .) Wegner draws our attention to the times—not infrequent among those of us who are "absent-minded"—when we find ourselves with a perfectly conscious thought that just baffles us; it is, as he wonderfully puts it, *conscious but not accessible* (Wegner 2002, p. 163). (Now why am I standing in the kitchen in front of the cupboard? I know I'm in the place I meant to be, but what did I come in here to get?) At such a moment, *I* have lost track of the context, and hence the *raison d'être,* of this very thought, this conscious experience, and so its meaning (and that's what is most important) is temporarily no more accessible to *me*—the larger me that does the policy-making—than it would be to any third party, any "outside" observer who came upon it. In fact, some onlooker might well be able to remind me of what it was I was up to. My capacity to be reminded (re-minded) is crucial, since it is only this that could convince me that this onlooker was right, that this was something *I* was doing. If the thought or project is anyone's, it is mine—it belongs to the me who set it in motion and provided the context in which this thought makes sense; it is just that the part of me that is baffled is temporarily unable to gain access to the other part of me that is the author of this thought.

I might say, in apology, that I was *not myself* when I made that mistake, or forgot what I was about, but this is not the severe disruption of self-control that is observed in schizophrenia, in which the patient's own thoughts are interpreted as alien voices. This is just the fleeting loss of contact that can disrupt a perfectly good plan. A lot of what *you* are, a lot of what you are doing and know about, springs from structures down there in the engine room, causing the action to happen. If a thought of yours is *only conscious*, but not also accessible *to that machinery* (to some of it, to the machinery that needs it), then *you* can't do anything with it and are left just silently mouthing the damn phrase to yourself, your isolated self, over and over. Isolated consciousness can indeed do nothing much on its own. Nor can it be responsible.

As Wegner notes, "If people will often forget tasks for the simple reason that the tasks have been completed, this signals a *loss of contact* [my italics—DCD] with their initial intentions once actions are over—and thus a susceptibility to revised intentions" (Wegner 2002, p. 167). A loss of contact between what and what? Between a Cartesian Self that "does nothing" and a brain that makes all the decisions? No. A loss of contact between the you that was in charge then and the you that is in charge now. A *person* has to be able to keep in contact with past and anticipated intentions, and one of the main roles of the brain's user-illusion of itself, which I call the self as a center of narrative gravity, is to provide *me* with a means of interfacing with myself at other times. As Wegner puts it, "Conscious will is particularly useful, then, as a guide to ourselves" (p. 328). The perspectival trick we need in order to escape the clutches of the Cartesian Theater is coming to see that *I,* the larger, temporally and spatially extended self, can control, to some degree, what goes on inside of the simplification barrier, where the decision-making happens, and that is why, as Wegner says, "Illusory or not, conscious will is the person's guide to his or her own moral responsibility for action" (p. 341).

I know that many people find it hard to grasp this idea or take it seriously. It seems to them to be a trick with mirrors, some kind of verbal sleight of hand that whisks consciousness, and the real Self, out of the picture just when it was about to be introduced. Echoing Robert

Wright, this view seems to many to deny the existence of consciousness instead of explaining how it came to exist. Where does consciousness come into the picture? It is already there, unnoticed in the activity just described. Mental contents become conscious not by entering some special chamber in the brain, not by being transduced into some privileged and mysterious medium, but by winning the competitions against other mental contents for domination in the control of behavior, and hence for achieving long-lasting effects—or as we misleadingly say, "entering into memory." And since we are talkers, and since talking to ourselves is one of our most influential activities, one of the most effective ways—not the only way—for a mental content to become influential is for it to get into position to drive the language-using parts of the controls. All this has to happen in the arena of the brain, in "central processing," but not under the direction of anything. As Ainslie noted, "The orderly internal marketplace pictured by conventional utility theory becomes a complicated free-for-all" (Ainslie 2001, p. 40), the Cartesian self fragmented into shifting coalitions, with no king or presiding judge.

> CONRAD: Suppose all these strange competitive processes are going on in my brain, and suppose that, as you say, the conscious processes are simply those that win the competitions. How does *that* make them conscious? What happens next to them that makes it true that *I* know about them? For after all, it is *my* consciousness, what I know from the first-person point of view, that needs explaining!

Such a question betrays a deep confusion, for it presupposes that what *you* are is something *else,* some Cartesian *res cogitans* in addition to all this brain-and-body activity. What you are, Conrad, just *is* this organization of all the competitive activity between a host of competences that your body has developed. You "automatically" know about these things going on in your body, because if you didn't, it wouldn't be your body!

The acts and events you can tell us about, and the reasons for them, are yours because you made them—and because they made you. What you are is that agent whose life you can tell about. You can tell us, and you can tell yourself. The process of self-description begins in earliest childhood, and includes a good deal of fantasy from the outset. (Think of Snoopy in the *Peanuts* cartoon, sitting on his doghouse and thinking, "Here's the World War I ace, flying into battle.") It continues through life. (Think of the café waiter in Jean Paul Sartre's discussion of "bad faith" in *Being and Nothingness* [1943], who is all wrapped up in learning how to live up to his self-description as a waiter.) It is what *we* do. It is what *we* are.[5]

The demands of communication don't just create the need for the sorts of self-monitoring arrangements that create the illusion of the Cartesian Theater. They also open up human psychology to a rich variety of further elaborations. The fact that the primary complexities in our environments are not just other agents—potential predators or prey or rivals or mates—but other *communicating* agents—potential friends or enemies, potential fellow citizens—has still further implications for the evolution of human freedom, to be drawn out in the remaining chapters.

Chapter 8

Exactly when and where do we make decisions? When we look closely at a person's conscious decisions, we discover that this quest for spatio-temporal precision breaks down, creating the illusion of an isolated, powerless self. We restore power, and hence the potential for moral responsibility, to the self by recognizing that its duties are distributed in both space and time in the brain.

Chapter 9

What are the prerequisites for autonomy and how could they ever be met? To be moral agents we must be capable of acting for reasons that are our reasons, but we are imperfect reasoners at best. Can we really be rational enough to sustain our sense of ourselves as genuine moral agents, and if so, how do we get that way?

---

5. Portions of the preceding three paragraphs are drawn, with revisions, from Dennett 1997B.

Notes on Sources and Further Reading

Libet's most recent essay on the topic appears in a volume inspired by his experiments, *The Volitional Brain* (Libet et al. 1999), that includes essays psychological, neurological, theological, philosophical, and just plain strange. This book is an unsurpassed paragon of open-mindedness, the proof of which is that it includes as its closing essay a trenchant review of itself, "A Review of *The Volitional Brain*," by Thomas Clark (1999), that incisively but fairly exposes all the major errors and many of the confusions in the essays that precede it. Philosophers have written quite a lot about the pragmatic contradiction involved in asserting such sentences as "*p* and nobody should believe that *p.*" Now they have a real-world example of that pragmatic contradiction on a large scale. (Actually, Stephen Stich beat them to it; the first chapter of his *Deconstructing the Mind* [1996] explicitly purports to refute the chapters that follow it: reprinted essays, some of them coauthored with graduate students of his. It is an example of changing one's mind in public that I wish more philosophers would emulate, though I do wonder about whether his various coauthors were quite as ready as he to abandon ship—they don't say.) My own discussions of Libet include Chapter 6, "Time and Experience," in *Consciousness Explained* (1991A); the somewhat more technical essay, written with Marcel Kinsbourne, "Time and the Observer: The Where and When of Consciousness in the Brain," *Behavioral and Brain Sciences,* 1991 (see also Libet's commentary in that volume); and my contributions, including debate with Libet, in the 1993 CIBA Foundation symposium volume, *Experimental and Theoretical Studies of Consciousness,* especially pp. 134–35. See also Libet's account in Libet 1996.

The philosophical literature on nefarious neurosurgeons implanting remote control devices in people's brains mostly grows out of the classic 1969 essay by Harry Frankfurt, "Alternative Possibilities and Moral Responsibility." See Kane 2001, and, best among recent books, John Martin Fischer and Mark Ravizza's *Responsibility and Control: A Theory of Moral Responsibility* (1998).

The permeable boundary between individual learning and "instinct" supported by genetic inheritance is opened up in particularly interesting ways by the Baldwin Effect, or what C. H. Waddington called genetic assimilation, a topic I discussed in both *Consciousness Explained* (Dennett 1991A) and *Darwin's Dangerous Idea* (Dennett 1995). A wave of second thoughts about the Baldwin Effect have recently been gathered in a forthcoming volume edited by Bruce Weber and David Depew, which includes an extended defense of the Baldwin Effect by me, "The Baldwin Effect: A Crane, not a Skyhook" (Dennett 2002B). I further developed other ideas in this chapter in *Kinds of Minds* (Dennett 1996A), "Learning and Labeling" (Dennett 1993), and "Making Tools for Thinking" (Dennett 2000A).

Chapter 9

BOOTSTRAPPING OURSELVES FREE

CULTURE

It is culture that has allowed us to become what Aristotle famously called us: rational animals. How? By permitting, once more, the division of labor and distribution of responsibility that has again and again achieved new levels of design sophistication in evolutionary history.

How We Captured Reasons and Made Them Our Own

We are creatures who ask why, with norms as in other domains. We want to take morality not blindly as a set of taboos but as something with a point—or perhaps more than one point, but then we want to think about what those points might have to do with each other and how to reconcile them.

—Allan Gibbard, *Wise Choices, Apt Feelings*

comm. beliefs/ plans

Human consciousness was made for sharing ideas. That is to say, the human user-interface was created by evolution, both biological and cultural, and it arose in response to a behavioral innovation: the activity of communicating beliefs and plans, and comparing notes. This turned many brains into many minds, and the distribution of authorship made possible by this interconnectedness is the source of not only our huge technological edge over the rest of nature but our morality. The last step required to complete my naturalistic account of free will and moral responsibility is to explain the R&D that has given us each a perspective on ourselves, a place from which to *take* responsibility. The name for this Archimedean perch is the self. This is something about us humans that sets us apart as potential moral agents, and it is no surprise that language is involved. What is harder to see is how language, when it is installed in a human brain, brings with it the construction of a new cognitive architecture that *creates* a new kind of consciousness—and morality.

This is both a historical question and a question of justification. If it were merely a historical question then the answer might be: Once upon a time, many years ago, space aliens came to Earth and made us all swallow morality pills; thereafter we taught morality to our children. Or only slightly more realistically: A retrovirus swept over our hominid ancestors, and in its wake the few who survived just happened to have a *gene for appreciating justice.* Or still more realistically: Morality memes arose by accident some tens of thousands of years ago and swept over the human population in a worldwide epidemic. Even if one of these fanciful tales were true, it would leave us without half the answer we require: What about justification?

Fortunately, Darwinian reasoning is all about explaining things "with a point." Any account in terms of natural selection presupposes an answer—one answer or another—to the question *Cui bono?* We will have to find further offspring of the Darwinian *Cui bono* questions, however, since the point of morality is manifestly *not* restricted to "the good of the species" or "the survival of our genes" or anything like that. It will have to be something that arises in the course of making ourselves the sort of selves that we are.

One of the disconcerting features of the evolutionary processes described in earlier chapters is the absence of anything like comprehension in the agents whose proclivities are shaped by the processes. These agents (or better, their genes) may be the beneficiaries of some amiable instincts, some bland *dispositions to cooperate,* but this is nothing to them. They can be oblivious to the reasons for the features that govern their lives, the free-floating rationales that they need not appreciate, and hence need not represent. The evolution of *our* capacity to recognize these reasons, and reflect on them, and thereby to change them into entirely different reasons, was another major transition in evolutionary history, and like all the others, it had to build on what had already evolved to serve other purposes.

The basic idea has been appreciated for centuries. According to David Hume, we begin with what he calls the natural motives: sexual appetite, affection for children, limited benevolence, interest, and

resentment—a list that any twenty-first-century evolutionary psychologist would look on with favor. These dispositions have rationales that
are not *our* rationales, though they have set the stage for our practice
of demanding and giving reasons. As Hume put it in his *Treatise of
Human Nature,* "If nature did not aid us in this particular 'twou'd be
in vain for politicians to talk of *honourable* and *dishonourable, praisewor-
thy* and *blameable.* These words wou'd be perfectly unintelligible"
(Hume 1739, p. 500). From the outset we find ourselves approving of
some attitudes and practices—as "intrinsically" good, somehow—and
these attitudes and practices were shaped over millennia without foresighted design but with their own *raisons d'être.* The benefits of some
of these entrenched habits and practices may have been at least dimly
perceived by our ancestors, but even this is not an exceptionless
requirement, since there are (at least) three ways in which differential
replication could pay for the designs bequeathed to us: (1) if our natural motives are adaptations directly advantageous to the individuals
who have them (individual-level selection, the more or less standard
case); (2) if there has been a group structure in human populations sufficiently salient to create conditions under which groups of unwitting
practice-followers could thrive at the expense of less favorably constituted groups (group selection), or (3) if memes for motives have been
in competition for the limited number of havens of human brains and
these motives have, like many of our other symbionts, gone to fixation
as stable features of human cultural ecology for one reason or another.
These are all "natural" ways, in Hume's sense, for us to be endowed
with the motives that provide the foundation for the next wave of
R&D, deliberate social engineering, which is only a few millennia old.
The natural motives, Hume maintained, have "offspring," which he
dubbed the "artificial" virtues of morality—such as justice. Hume saw
ethics as a kind of human technology, and he saw reflection as the tool
we get from nature that permits us to revise our natural instincts,
enhancing them with prosthetic elaborations whose rationales (free-
floating until Hume and others pinned them down and represented
them) are, in fact, aimed at still more freedom, consistent with security from harm. Eyeglasses for the soul, you might say. But before we
turn to this new sort of R&D, we should consider in outline the sort
of evolutionary process made possible by the transition from oblivious
agents to minded, reflective agents.

We start with Brian Skyrms's elegant "evolutionary fable" of the cake-splitting game in his book *Evolution of the Social Contract* (1996, pp. 3ff). Suppose you and I come upon a chocolate cake we want to divide between us. Instead of fighting over it (a dangerous option for both of us) we agree to settle the matter by a simple game: "Each of us writes a final claim to a percentage of the cake on a piece of paper, folds it, and hands it to a referee. If the claims total more than 100%, the referee eats the cake. Otherwise we get what we claim. (We may suppose that if we claim less than 100% the referee gets the difference.)" (p. 4). As Skyrms notes, almost everyone would choose 50%, the fair amount. (The referee is not really part of the model, but just a bit of stage-dressing.) And sure enough, evolutionary game theory shows that the 50–50 split is an *evolutionarily stable strategy,* or ESS. "Fair division will be stable in any dynamics with a tendency to increase the proportion (or probability) of strategies with greater payoffs because any unilateral deviation from fair division results in a strictly worse payoff" (p. 11). But it is not the only ESS, Skyrms notes; there are many others. This is the problem of *polymorphic traps:*

> For example, suppose that half the population claims ⅔ of the cake and half the population claims ⅓. Let us call the first strategy *Greedy* and the second *Modest*. A greedy individual stands an equal chance of meeting another greedy or a modest individual [because we have not yet introduced any correlation—DCD]. If she meets another greedy individual she gets nothing because their claims exceed the whole cake, but if she meets a modest individual, she gets ⅔. Her average payoff is ⅓. A modest individual, on the other hand, gets a payoff of ⅓ no matter who she meets.
>
> Let us check to see if this polymorphism is a stable equilibrium. First note that if the proportion of greedys should rise, then greedys would meet each other more often and the average payoff to greedy would fall below the ⅓ guaranteed to modest. And if the proportion of greedys should fall, the greedys would meet modests more often, and the average payoff to greedys would rise above ⅓. Negative feedback will keep the population proportions of greedy and modest at equality. But what of the invasion of other mutant strategies? Suppose that a

Supergreedy mutant who demands more than ⅔ arises in the population. This mutant gets payoff of 0 and goes extinct. Suppose that a *Supermodest* mutant who demands less than ⅓ arises in the population. This mutant will get what she asks for, which is less than greedy and modest get, so she will also go extinct—although more slowly than supergreedy will. The remaining possibility is that a middle-of-the-road mutant arises who asks for more than modest but less than greedy. A case of special interest is that of the *Fair-minded* mutant who asks for exactly ½. All of these mutants would get nothing when they meet greedy and get less than greedy does when they meet modest. Thus they will all have an average payoff less than ⅓ and all—including our fair-minded mutant—will be driven to extinction. The polymorphism has strong stability properties.

This is unhappy news, for the population as well as for the evolution of justice, because our polymorphism is inefficient. Here everyone gets, on average, ⅓ of the cake—while ⅓ of the cake is squandered in greedy encounters. (Skyrms 1996, pp. 12–13)

Skyrms goes on to note that once we add some positive correlation to the picture, so that each type of strategy tends to interact with its own kind more than random pairing would ensure, these unfortunate polymorphisms become less attractive—they become more evitable. It doesn't matter what feature of the world makes this correlation increase, but *agents with minds and culture* are particularly well-suited to achieve this, as Don Ross demonstrates in an imaginative Just So Story that builds on Skyrms.

Imagine a population that's settled into one of the polymorphic ESSs. Continued success for the Greedy agents in this game will depend on encouraging Modest agents to avoid interaction with any Fairman mutants who come along. So we'd expect this population to evolve norms of justice that are a bit like Aristotle's. These norms will associate "justice" with the idea that the Modest should respect their natural station and defer to the Greedy. These will be norms very familiar from many human societies, past and present. If these agents can't do moderately sophisticated computations, or pass around their thoughts on the implications of these, there the population will stay. It is, after all, in ESS equilibrium. But if these agents can do just a little economics and *also* grasp basic Darwinian logic—nothing very fancy is needed—they can notice that the all-Fairman ESS is (a) more efficient (the economic point), and (b) achievable along an equilibrium path (the Darwinian point). We can readily imagine what might happen. Initially, most of the population will find the idea of the all-Fairman ESS a shocking violation of natural morality. But a *few* of the Modests will get from recognition of (a) to the concept of their own exploitation. Why not? Any creature with much conceptual flexibility will try that reasoning step, even if only to talk himself out of the conclusion in deference to public opinion. Some Modests who embrace the idea will be persecuted; but this will itself help spread the meme by dramatizing its importance. The enlightened Modests, if only they can identify each other, can easily rebel in a quietly effective way: they just need to play the Fairman strategy with each other, thereby realizing the greater gains from trade. After all, when we talk of "Fairman mutants" arising here we needn't literally mean genetic freaks; every time the Fairman meme lodges in a Modest's mind, we have a mutant. So far, let us suppose, these mutants are just motivated by acquisitiveness: they haven't yet *morally* challenged the prevailing norms. Some Modests, and even some Greedies, though, will find the mathematical beauty of the more efficient outcomes attractive enough to strive for in their own right. This will complement self-interest in speeding up the dynamic, though it isn't strictly necessary.

Evolutionary game theory shows that this population will evolve relentlessly toward the all-Fairman ESS. Well before that point, the concept of justice-as-fairness will *naturally* arise, because the Fairmen best promote their own success by encouraging ostracism of the Greedy. Inculcating moral revulsion against Greedy strategies will be a natural move—a very obvious Good Trick—if they're biologically equipped to experience *simple* revulsion against anything. Eventually, the population will look back on its own former consensus (if they're sophisticated) as a kind of amoral childishness. If they're not sophisticated, they'll decide that their ancestors were bad, and some silly and insecure ones will discourage the reading of their surviving books.

Now, look what's happened here. These agents underwent moral evolution, measurable as such by an objective standard. They got there by, at step one, catching a glimpse of an elementary Darwinian logical point. No farsighted moral superhero, neither Christ nor Nietzsche, had to exhort them along at any point. A bit of science and logic did the whole trick. At the end of the process, do these agents know something their ancestors didn't? Sure they do: they know that fairness is just; they really *are* morally superior to their ancestors. Hume's Guillotine [the principle that you can't derive "ought" from "is"—DCD] notwithstanding, they found this out thanks to being conscious meme-spreaders who could think about hypotheticals, and thanks to using those capacities to learn some evolutionary theory. (Ross, personal correspondence)

Of course, you don't have to use the language of professional economics to appreciate the economic point, and you don't have to be an explicit Darwinian to see how you might get from here (inefficient polymorphic trap) to there (fair distribution) by a self-sustaining path. A semi-understood, dimly imagined version will do just fine, as always, as we pick our way gradually from obliviousness to comprehension. Darwin himself drew our attention to the importance of what he called *unconscious selection* as an intermediate step between *natural selection* and what he called *methodical selection*: the deliberate, foresighted, intended "improvement of the breed" by animal and plant breeders. Darwin pointed out that the line between unconscious and methodical selection was itself a fuzzy, gradual boundary:

> The man who first selected a pigeon with a slightly larger tail, never dreamed what the descendants of that pigeon would become through long-continued, partly unconscious and partly methodical selection. (Darwin 1859, p. 39)

And both unconscious and methodical selection are just special cases of the more inclusive process, natural selection, in which the role of human intelligence and choice can stand at zero. From the perspective of natural selection, changes in lineages due to unconscious or methodical selection are merely changes in which one of the most prominent selection pressures in the environment is human activity. This nesting of different processes of natural selection of genes has more recently spawned a new member: genetic engineering. How does it differ from the methodical selection of Darwin's day? It is less dependent on the preexisting variation in the gene pool, and proceeds more directly to new candidate genomes, with less overt and time-consuming trial and error. There is ever more accurate foresight, but even here, if we look closely at the practices in the laboratory, we will find a large measure of exploratory trial and error in their search of the best combinations of genes.

We can use Darwin's three levels of genetic selection, plus our more recent fourth level, genetic engineering, as a model for four parallel levels of *memetic* selection in human culture. The first memes were naturally selected, paving the way for unconsciously selected memes—memes that were "domesticated" by inadvertence, we might say—followed by methodically selected memes, in which human foresight and planning played a clear role, but in which the underlying mechanisms were only dimly understood and most experimentation consisted of seeking simple variations on the existing themes, to the present day when memetic engineering is a major human enterprise: the attempt to design and spread whole systems of human culture, ethical theories, political ideologies, systems of justice and government, a cornucopia of competing designs for living in social groups. Memetic engineering is a very recent sophistication in the history of evolution on this planet, but it is still several millennia older than genetic engineering; among its first well-known products are Plato's *Republic* and Aristotle's *Politics*.

We are not just Popperian creatures, capable of thinking ahead and imagining alternative futures and their likely outcomes, but Gregorian creatures, using the thinking tools that our cultures install in us during childhood and beyond (Dennett 1995, pp. 377 ff.). We come to share a grab bag of memorized precepts on the tips of our tongues as we face life's dilemmas. Even fairy tales and Aesop's fables have a valuable role to play in directing a child's attention. One of the reasons we so seldom paint ourselves into a corner or saw off the limb we are sitting on is that we have all heard one funny, memorable tale or another about a chap who did just that. And if we follow the Golden Rule, or the Ten Commandments, we are enhancing our underlying natural instincts with prosthetic devices that tend to encourage framing the situations we confront in one way or another. But much of this lore has itself evolved without deliberate authorship, and has been passed on without explicit appreciation of its utility, until quite recently.

Psychic Engineering and the Arms Race of Rationality

In effect I took the standpoint of a psychic engineer charged with designing our norms for an advantage we recognize together.

—Allan Gibbard, *Wise Choices, Apt Feelings*

Once we have captured the free-floating rationales of the natural motives and represented them, alongside the representations of all the artifices we dream up in the course of our reflections, we are no longer bound by the inefficient, wasteful, mindless trial and error of natural selection. We can hope to replace an equilibrium of sheer replicative power with a *reflective equilibrium* of rational agents who have engaged in the communal activity of mutual persuasion. This shift from un-directed trial and error to intelligent (re-)design is, I have suggested, a major transition in evolutionary history, opening up literally undreamed of dimensions of opportunity, for good or ill. Until the birth of ethics, Darwinian R&D had proceeded for billions of years without any foresight, gradually climbing the slopes of Mount Improb-able (Dawkins 1996). Wherever lineages found themselves on local peaks of the adaptive landscape, their members had no way of so much as wondering whether or not there might be higher, better summits on the far side of this valley or that. In their physical landscapes, the more farsighted of them could do something tantamount to framing the goal of getting to the other side of the river, or to the visible patch of edible grass on that hilltop over there, but more remote questions about what the point of living might be and how it might best be achieved were inexpressible until we came along. We are the only species whose members can *imagine* the adaptive landscape of possibil-ities beyond the physical landscape, who can "see" across the valleys to other conceivable peaks. The mere fact that we're doing what we're doing—trying to figure out whether our ethical aspirations have any sound anchoring in the world science is uncovering for us—shows how different we are from all other species.

by our own ends & meant [handwritten marginalia]

We can conceive (we think) of better worlds and yearn to get there. Are we right that these other worlds might be better? In what sense? By whose standards? By ours. Our evolved capacity to reflect gives us—and only us—both the opportunity and the competence to evaluate the ends, not just the means. We have to use our current values as the starting point for any contemplated revaluation of values, but from our perspective on our current hilltop, we can formulate, criticize, revise, and—if we are lucky—mutually endorse a set of design principles for living in society. We can envisage some tempting utopian peaks quite different from our current circumstances. Can we get to any of them? Are we sure we want to try? If we can't get there, that may be tragic, but not an offense against reason. How to factor in politics, the art of the possible, is itself one of the most difficult design questions we face. We may be stuck, alas, in the best of all possible worlds, given our historical predicament, but then again, we may be able to discover some adjustments in our current design that have some hope of carrying us to higher summits. And unlike all other species, these are problems *for us.* We actually work on them, devoting time and energy to them. We gather information relevant to them, explore variations on them, and debate their merits knowing that our reflections will actually help determine which trajectory our future holds.

design principles for society [handwritten marginalia]

This provides, at last, a naturalistic framework within which the traditional questions of morality can make sense. Our evolutionary journey has brought us to the traditional arena of philosophical and political investigation and debate, in which many ideas compete for our endorsement. Ethics is a large and complex field, and this is a contest I will make no attempt to adjudicate or even contribute to in this book, beyond a few suggestions about some fossil traces of that journey that can still deflect our ethical thinking in misleading ways. One of our most pressing tasks, as psychic engineers, is to see if we can secure the fundamental concept of a responsible moral agent, an agent who, unlike the cooperative prairie dog or loyal wolf or friendly dolphin, chooses freely for considered reasons and may be held morally accountable for the acts chosen. We have sketched the evolutionary development of the patterns that constitute the conceptual environment in which such a concept can reside—the air we breathe—but we also need to examine more closely how an individual might be able to grow into such an exalted role. Can anybody actually make the grade? Aren't

morally accountable agent [handwritten marginalia]

we learning from psychologists that we are *actually* a far cry from the rational agents we pretend to be?

Allen Funt was one of the great psychologists of the twentieth century. His informal experiments and demonstrations on *Candid Camera* showed us as much about human psychology and its surprising limitations as the work of any academic psychologist. Here is one of the best (as I recall it many years later): He placed an umbrella stand in a prominent place in a department store and filled it with shiny new golf-cart handles. These were pieces of strong, gleaming stainless steel tubing, about two feet long, with a gentle bend in the middle, threaded at one end (to screw into a threaded socket on your golf cart) and with a sturdy spherical plastic knob on the other end. In other words, about as useless a piece of stainless steel tubing as you could imagine, unless you happened to own a golf cart missing its handle. He put up a sign. It didn't identify the contents but simply said: "50% off. Today only! $5.95." Some people purchased them, and when asked why, they were quite ready to volunteer one confabulated answer or another. They had no idea what the thing was, but it was a handsome thing, and such a bargain! These people were not brain-damaged or drunk; they were normal adults, our neighbors, ourselves.

We laugh nervously as we peer into the abyss that such a demonstration opens up. We may be smart, but none of us is perfect, and whereas you and I might not fall for the old golf-cart-handle trick, we know for certain that there are variations on this trick that we have fallen for and no doubt will fall for in the future. When we discover our imperfect rationality, our susceptibility to being moved in the space of reasons by something other than consciously appreciated reasons, we fear that we aren't free after all. Perhaps we're kidding ourselves. Perhaps our approximation of a perfect Kantian faculty of practical reason falls so far short that our proud self-identification as moral agents is a delusion of grandeur. failures of freedom '

Our failures in such cases are indeed failures of freedom, failures to respond as we would want to respond to the opportunities and crises life throws at us. For that reason they are ominous, for this is indeed one of the varieties of free will worth wanting. Notice that Funt's demonstration would not impress us if his subjects were not people but animals—dogs or wolves or dolphins or apes. That a mere beast can be tricked into opting for something shiny and alluring but not what the beast truly wants—should truly want—is hardly news to us. We expect "lesser" animals to live in the world of appearances, beneficiaries of "instincts" and perceptual capacities that are magnificently effective in context, but easily exposed in unlikely circumstances. We aspire to a higher ideal.

As we learn more and more about human weaknesses and the way the technologies of persuasion can exploit them, it can seem as if our vaunted autonomy is an unsupportable myth. "Pick a card, any card," says the magician, and deftly gets you to pick the card he has chosen for you. Salespeople know a hundred ways to get you off the fence so that you buy that car, that dress. Lowering one's voice, it turns out, works very well: "*I see you in the green number.*" (You might want to remember that the next time a salesperson whispers at you.) Notice that there is an arms race here, with ploy and counterploy balancing each other out. I've just somewhat diminished the effectiveness of the whispering trick against those of you who remember my exposure of it. It is easy enough to discern the ideal of rationality that serves as the background for this battle: *Caveat emptor,* we declare, let the buyer beware. This is a policy that presupposes that the buyer is rational enough to see through the blandishments of the seller, but since we know better than to believe this myth taken neat, we go on to endorse a policy of *informed consent,* prescribing the explicit representation in clear language of all the relevant conditions for one agreement or another. Then we also recognize that such policies are subject to extensive evasion—the fine-print ploy, the impressive-sounding gobbledygook—so we may go on to prescribe still further exercises in spoon-feeding the information to the hapless consumer. At what point have we abandoned the myth of "consenting adults" in our "infantalizing" of the citizenry? When we learn of proposals to tailor a message to particular groups or particular individuals, each group targeted with specific images, stories, aids, and warnings, we may be tempted to condemn these proposals as paternalistic, and as subversive to the ideal of free will in which we are Kantian rational agents, responsible for our own destiny. But at the same time we should acknowledge that the environment we live in has been being updated ever since the dawn of civilization, elaborately prepared, made easy for us, with multiple signposts and alerts along the way, to ease the burdens on us imperfect decision-makers. We happily lean on the prostheses that *we* find valuable—that's the beauty of civilized life—but

arms race— not just one or other

tend to begrudge those that others need. Once we understand that this is an arms race, we can fend off the absolutism that sees only two possibilities: Either we are perfectly rational or we are not rational at all. That absolutism fosters the paranoid fear that science might be on the verge of showing us that our rationality is an illusion, however benign from some perspectives. That fear, in turn, lends spurious attractiveness to any doctrine that promises to keep science at bay, our minds sacrosanct and mysterious. We are actually wonderfully rational. We are rational enough, for instance, to be really good at designing ploys for playing mind games on each other, seeking out ever more subtle chinks in our rational defenses, a game of hide-and-seek with no time out or time limit.

But how do we get good enough at this to make the team? A good answer to this question must fend off paradox on all sides (Suber 1992). If you are free, are you responsible for being free, or just lucky? Can you be blamed for failing to make yourself free? As we saw in Chapter 7, cooperators capable of solving commitment problems and establishing their reputation as moral agents get to enjoy the many benefits of being a trusted member of the community, but if you have not yet achieved that status, what hope, if any, is there for you? Should we regard the frequent defectors among us with contempt or compassion? Boundaries created by evolutionary processes tend to be porous and gradual, with intermediate cases bridging the chasms between the Haves and the Have-nots, but we cannot go along with Mother Nature's refusal to categorize all the way down. Our moral and political systems apparently oblige us to sort people into two categories: those who are morally responsible and those who are excused because they don't make the grade. Only the former are fit candidates for punishment, for being held accountable for their misdeeds. How shall we decide where to draw the line? The occasionally stupid actions we take and the habits and character traits we discover in ourselves may make us wonder whether any such categorization can be anything but a convenient myth, rather like Plato's odious myth of the metals, the pioneering public relations ploy by which he proposed to keep the peace in his Republic. Some people are born to be Gold, and others should be content to be Silver or Bronze. It may seem, for instance, that political theory endorses the policy of maintaining a certain degree of punishment in a society, in order to render credible the prohibitions that

2 Categories

[handwritten: punishment]

actually deter the rational from transgressing (to some degree), but this policy is doomed to hypocrisy. Those whom we end up punishing are really paying a double price, for they are scapegoats, deliberately harmed by society in order to set a vivid example for the more ably self-controlled, but not really responsible for the deeds we piously declare them to have committed of their own free will. What, in fact, are the qualifications for being a genuinely culpable miscreant, and could anybody actually meet them?

With a Little Help from My Friends

The things of which romance assures him are very far from being true: yet it is solely by believing himself a creature but little lower than the cherubim that man has by interminable small degrees become, upon the whole, distinctly superior to the chimpanzee.

—James Branch Cabell, *Beyond Life*

Fake it until you make it.

—a slogan of Alcoholics Anonymous

In Chapter 4, we considered and rejected Robert Kane's attempt to halt a threatened infinite regress with some rather magic moments—Self-Forming-Actions, or SFAs—buck-stopping instants in which the universe holds its breath while a quantum indeterminacy permits you to "do-it-yourself," creating yourself as a responsible moral agent (and you could have done otherwise). Kane's solution won't work because you can't stop a regress by invoking a Prime Mammal, by inventing a special difference that is "essential" but invisible. A *genuine* quantum-choice person and her pseudo-random-choice twin, like the Prime Mammal and the Prime Mammal's mother, differ in no discernible regard that could make such a special difference. You could never tell you'd succeeded in having a genuine SFA, so even if they do occur, their moral significance evaporates on examination and the regress still threatens. How, then, if not by a miraculous leap of self-creation, did you get here (moral agency) from there (the amoral unfreedom of an infant)? Not surprisingly, my answer will invoke the Darwinian themes

[handwritten: don't know if you have SFA]

of luck, environmental scaffolding, and gradualism. With a little bit of luck, and a little help from your friends, you put your considerable native talent to work, and bootstrapped your way to moral agency, inch by inch.

The basic process was outlined in Chapter 8: A proper human self is the largely unwitting creation of an interpersonal design process in which we encourage small children to become communicators and, in particular, to join our practice of asking for and giving reasons, and then reasoning about what to do and why. For this to work, you have to start with the right raw materials. You won't succeed if you try it with your dog, for instance, or even a chimpanzee, as we know from a series of protracted and enthusiastic attempts over the years. Some human infants are also unable to rise to the occasion. The first threshold on the path to personhood, then, is simply whether or not one's caregivers succeed in kindling a communicator. Those whose fires of reason just won't light for one reason or another are consigned to a lower status, uncontroversially. It's not their fault, it's just their bad luck. But while we're on the topic of luck, let's first try to calibrate our scales. Every living thing is, from the cosmic perspective, incredibly lucky simply to be alive. Most, 90 percent and more, of all the organisms that have ever lived have died without viable offspring, but not a single one of your ancestors, going back to the dawn of life on Earth, suffered that normal misfortune. You spring from an unbroken line of winners going back billions of generations, and those winners were, in every generation, the luckiest of the lucky, one out of a hundred or a thousand or even a million. So however unlucky you may be on some occasion today, your presence on the planet testifies to the role luck has played in your past.

Above the first threshold, people exhibit a wide diversity of further talents, for thinking and talking, and for self-control. Some of this difference is "genetic"—due mainly to differences in the particular set of genes that compose their genomes—and some of it is congenital but not directly genetic (due to their mother's malnutrition or drug addiction, or to fetal alcohol syndrome, for instance), and some of it has no cause at all, in the sense we discovered in Chapter 3: It is the result of chance. None of these differences in your legacy are factors within your control, of course, since they were in place before you were born. And it is true that the foreseeable effects of some of them are inevitable, but not all—and less and less each year. It is also not in any way your own doing that you were born into a specific milieu, rich or poor, pampered or abused, given a head start or held back at the starting line. And these differences, which are striking, are also diverse in their effects—some inevitable and some evitable, some leaving life-long scars and others evanescent in effect. Many of the differences that survive are, in any event, of negligible importance to what concerns us here: a second threshold, the threshold of moral responsibility—as contrasted, say, with artistic genius. Not everybody can be a Shakespeare or a Bach, but almost everybody can learn to read and write well enough to become an informed citizen.

When W. T. Greenough and F. R. Volkmar (1972) first demonstrated that rats given a rich environment of toys and exercise gear and opportunities for vigorous exploration had measurably more neural connections, and larger brains, than rats raised in a bare, restrictive environment, some parents and educators went overboard in their eagerness to herald this important discovery, and then began to worry themselves sick over whether junior was getting enough of the right kinds of crib toys. In fact, we've known forever that a child raised alone in a bare room with no toys at all will be seriously stunted, and nobody has yet shown that the difference between having two toys and having twenty toys or two hundred toys makes any noticeable long-term difference in how the infant's brain develops. It would be extremely hard to show because so many confounding intervening influences, some planned and some fortuitous, would do and undo the effect that concerns us a hundred times a year as each child matured. We should do the difficult research as best we can, since it is *possible* that one condition or another is playing a larger role than suspected—and hence is a more appropriate target at which to aim our efforts of avoidance. But we can already be quite sure that most, if not all, of these differences in starting conditions vanish into the statistical fog as time passes. Like coin tosses, there may be no salient causation to be discerned in the outcomes. Once we have disentangled these factors to the extent that this is possible with careful scientific study, we will be able to say with some deserved confidence which interventions are apt to be needed in order to counteract which shortcomings, and only then will we be in a good position to make the value judgments that everybody is aching to make.

Tom Wolfe, for instance, deplores the use of Ritalin (methylphenidate) and other methamphetamines to counteract attention deficit hyperactivity disorder in children. He does this without pausing to consider the mass of evidence that indicates that *some* children have a readily correctable—evitable—dopamine imbalance in their brains that gives them a handicap in the self-control department just as surely as myopia does in the vision department.

> An entire generation of American boys, from the best private schools of the Northeast to the worst sludge-trap public schools of Los Angeles and San Diego, was now strung out on methylphenidate, diligently doled out to them every day by their connection, the school nurse. America is a wonderful country! I mean it! No honest writer would challenge that statement! The human comedy never runs out of material! It never lets you down!
>
> Meantime, the notion of a self—a self who exercises self-discipline, postpones gratification, curbs the sexual appetite, stops short of aggression and criminal behavior—a self who can become more intelligent and lift itself to the very peaks of life by its own bootstraps through study, practice, perseverance, and refusal to give up in the face of great odds—this old-fashioned notion (what's a *boot*strap, for God's sake?) of success through enterprise and true grit is already slipping away, slipping away . . . slipping away. (Wolfe 2000, p. 104)

This characteristically purple passage has some uncharacteristically unintended irony in it. I wonder if Wolfe would commend a bracing regimen of eye exercises and courses in Learning to Live with Short-Sightedness in lieu of eyeglasses for the myopic. He ends up declaiming the twenty-first-century version of that old chestnut: If God had meant us to fly, he would have given us wings. So rattled is he by the imaginary bogey of genetic determinism that he cannot see that the bootstrapping he yearns to protect, the very fount of our freedoms, is enhanced, not threatened, by demythologizing the self. Scientific knowledge is the royal road—the only road—to evitability. Perhaps here we see the outlines of a secret fear that lies behind some of the calls to *Stop that crow!* Not that science will take away our freedom, but that science will give us too much freedom. If your child doesn't have as much "true grit" as your neighbor's child, perhaps you can buy him some artificial grit. Why not? It's a free country, and self-improvement is one of our highest ideals. Why should it be important that you do all your self-improvement the old-fashioned way? These are very important questions, and their answers are not obvious. They should be addressed directly, not distorted by an ill-advised attempt to smother them.

In *Elbow Room,* I compared differences in initial endowment, genetic and environmental, to the staggered start of a marathon, in which some runners start many meters behind others but all head for the same finish line. This is fair, I argued, since in such a long race, "such a relatively small initial advantage would count for nothing, since one can reliably expect other fortuitous breaks to have even greater effects" (Dennett 1984, p. 95). This is true, but it underplays the role of *non*-fortuitous breaks in the race to responsible agenthood. The quest for personhood is something of a team effort, with coaches and supporters playing important roles on the sidelines, enriching the environment with a kind of scaffolding designed (unconsciously) to bring out the best in us. More important than the supply of developmentally appropriate toys, and even proper nutrition, is the set of ambient attitudes and policies a child observes and eventually participates in. There is a body of evidence supporting the hypothesis that children exposed to others who are violent, lying, and uncaring—playmates as much or even more than parents—tend to perpetuate those character traits. The silver lining of this cloud is just as important: Those of us who have the good fortune to grow up in free societies, in the company of people who are reasonable, truthful, and loving, tend to aspire to *those* ideals. Upbringing does make a big difference.

It is a mistake to reduce the effects of upbringing to "moral education," as if the key to ensuring that one's wards grow up into responsible adults was a dutiful attention to one catechism or another. Having a *vade mecum* of pithy precepts in one's kit is helpful, but a more potent set of influences is installed even earlier, and it channels the way we think every fleeting thought. When we talk to our pre-linguistic children, we half-consciously know that most of what we are saying to them goes over their heads, but not all. Some of it sticks. What do you *want?* Are you *afraid* of that? Where does it *hurt?* Do you *know* where the bunny is? Are you *trying* to *fool* me? "Don't worry, she'll grow into it," Mother says, when imposing somewhat outsize hand-me-down clothes on her child, and much the same could be said about the somewhat outsize hand-me-down psychological dispositions adults impose on us when we are children. Sure enough, we do grow into them, making them ours, making us them, making us into agents like the grown-ups. The more seriously we take our children as participants in the practice of asking and giving reasons, the more seriously they will come to take themselves.

This penchant for *presuming,* for erring on the side of presupposing more design competence in our young interlocutors than the cold facts might warrant, is an extraordinarily powerful addition to the Darwinian arsenal of R&D tricks. Because we human beings are not blind watchmakers, but sighted selfmakers, who moreover can reflect on what we see and draw inferences about what we wish to see in the future, we are much more readily redesigned, first by others, and then by ourselves, than any other organism yet evolved on the planet. Consider, for instance, the phenomenon of "being on one's best behavior." Independently of all instruction, formal or informal, we almost always adjust our behavior to harmonize with (what we take to be) the social demands of the current circumstance. Aside from a few bizarrely free spirits who seem genuinely unmoved by social pressure, people find that it is only with the most strenuous and disciplined of efforts that they can deliberately thwart the expectations of those around them. This pressure of expectation works in all directions. What parent hasn't discovered new strengths of character, new triumphs over sloth or fear or squeamishness, in the recognition that one's child is watching? Since we "rise to the occasion," it is a good thing to have a life full of occasions, opportunities to display our better selves, to others *and to ourselves,* and thereby to make it more likely that these better selves will make an easier appearance in the future. (Ainslie 2001 has a particularly insightful discussion of this dynamic.) The "presentation of self in everyday life" (Goffman 1959) is an elaborately (but mostly unconsciously) choreographed interactive dance, in which we not only attempt to appear better than we are, but in the process bring out the best in others. One wouldn't want to tamper carelessly with that set of practices, the fruits of thousands of years of genetic and cultural evolution. It could undo a lot of valuable R&D. *(Stop that crow!)* On the other hand, if done with discernment and understanding, some tampering could strengthen and enhance those designs, making up for

missed opportunities or blurred perceptions. Moreover, some deliberate intervention might help extinguish any unfortunate variants of our
practices that can be seen to be self-defeating. This is where our
evolved capacity for reflection comes into its own. Consider the
subtle but devastating effect discerned by the African-American writer
Debra Dickerson, writing about her father:

> Later, I came to understand that he both expected and needed
> blacks to fail, otherwise there was no proof of white perfidy and
> soullessness. He never understood that his fatalism was a self-
> fulfilling, self-defeating prophecy. He never considered that he
> had to believe, at some level, that whites were superior since he
> believed blacks had no chance whatsoever in life—but probably,
> he would have attributed that to the transcendent power of the
> white's innate evil. Among ourselves, we say "the white man's ice
> is colder" to describe the many of us who won't believe or value
> anything unless it comes from white people. The worse off some
> blacks are, the more magical whites seem, albeit an evil magic.
>
> So my father, like many other blacks, did the oppressor's job
> for him; he taught me to do the same. This was the moment that
> I began to close doors on myself. Perhaps whites would have been
> happy to take that task on themselves, but they rarely had to.
> Whites didn't have to place barriers in my path, I did it myself by
> "accepting" my preordained place at the end of every line.
> Racism and systematic inequality are very real forces in all our
> lives, but so is fatalism and a perverse kind of exultation of oppres-
> sion. (Dickerson 2000, p. 40)

What are the larger-scale social patterns that will enhance freedom and distribute it more equitably across the planet? What combination of explicit promulgations and subtle tricks is most likely to
flavor the environment in ways conducive to the growth of human
selves? In Chapter 7, we considered Robert Frank's suggestion that
problems of self-control and commitment help solve each other by
favoring the evolution of such emotions as anger and love. Allan Gibbard expands the point by addressing the issue of how a "psychic engineer" might want to fine-tune people's dispositions to feel anger, guilt,
and other emotions. Anger, Gibbard notes, "is powerful and inevitable,
and it often helps regulate action in desirable ways" (Gibbard 1990,

p. 298). Although "we are stuck with anger, whatever our norms" (p. 299), some cultures seem not to have any role for guilt. This raises the question of whether we might all be better off without it. Some hard determinists have argued that we not only should not bewail the passing of "genuine" free will; we should say good riddance, since without the presumption of free will, we may abandon the presumptions of moral responsibility, blame, and retribution, and all live more happily ever after. I have done what I can to sever the connection they imagine between determinism and responsibility, but still we can consider, with Gibbard, whether or not morality itself is a feature we should try to preserve in our societies. "In part the question is pragmatic: might we do best without these particular feelings, or without norms to govern them?" (p. 295). Guilt and anger mesh together well: Guilt placates anger, and the threat of guilt averts acts that would evoke anger. How would people tend to behave toward each other in a society in which guilt and anger were both damped down as far as possible, or—with heroic re-engineering of society—obtunded altogether? Might it even be wise, for one reason or another, to tune guilt and anger out of balance, with one of them overshooting a bit? The hard determinists say that our world would be a better place if we could somehow talk ourselves out of feeling guilty when we cause harm and angry when harm is done to us. But it isn't clear that any feasible "cure" along these lines wouldn't be worse than the "disease." Anger and guilt have their rationales, and they are deeply embedded in our psychology.

Better, Gibbard argues, is a policy that favors conditions that will moderate the intensity of the norms for these emotions. He contrasts "imperious" with "diffident" designs of moral norms. Imperious norms demand a lot, and hence foster private reservations, hypocrisy, and suspicion of others. They put a severe strain on human nature and tend to involve "somewhat inefficient hectoring." This, he claims, is a straightforward design flaw, like setting the steering ratio on a car too high, leading drivers to oversteer, overcorrect, overcorrect their corrections, and so on. It is unsafe, and puts undue strain on the mechanism, without achieving its intended effects (p. 306). Diffident norms, on the other hand, are relatively easygoing, a compromise with prudence and self-interest that is easier to swallow, and hence easier for individuals actually to endorse. So Gibbard suggests that the rational designer will tune the norms for anger and guilt

egoistic goals vs. morality

rather diffidently, a culturally inculcated setting that harnesses nature instead of fighting it.

Consider an individual Gibbard calls the "private ruminator," who is faced with a competition between egoistic goals and the tug of general benevolence, or morality. He is driven in public discussion to voice agreement with various publicly endorsed norms but may have private reservations, may ask himself if he really should go along with them when he can get away with not doing so. He may be familiar with Robert Frank's claim that it pays, prudentially, to *be* good in order to *seem* good, but he may toy with the idea that he's the exception. He has accepted help from his friends but is capable of wondering just how good a bargain this is, if it requires him to reciprocate in helping them. Has he been cajoled into good citizenship by the situational demands of conversation? How this conflict is resolved may depend heavily on the societal atmosphere:

> If allegiance to morality is the best way to promote his more egoistic goals, then his ambivalence is resolved. With an imperious morality this is unlikely; with diffident morality, it seems more plausible. . . . What makes a morality diffident is that it allies itself with enough other motives to prevail, for the most part—to prevail with actual people, with all their jointness and separateness, with their normative motivations and their appetites, feelings, impulses, and yearnings. (Gibbard 1990, p. 309)

Engineers, like politicians, are concerned with the art of the possible, and this requires us, above all, to think realistically about what people actually are, and how they got that way. Exercises in ethical theorizing that refuse to bow to the empirical facts about the human predicament are bound to generate fantasies that may have some aesthetic interest but ought not to be taken seriously as practical recommendations. Like everything else evolution has created, we're a somewhat opportunistically contrived bag of tricks, and our morality should be based on that realization. Philosophers have often attempted to establish a hyperpure, ultra-rational morality, untainted by "sympathy" (Kant) or "instinct," by animal dispositions or passions or emotions at all. Gibbard looks pragmatically at what we have to work with and proposes to do, as an engineer, what Mother Nature has always done: work with what you have.

Work w/ what you have

Autonomy, Brainwashing, and Education

To take oneself as a rational agent is to assume that one's reason has a practical application or, equivalently, that one has a will. Moreover, one cannot assume this without already presupposing the idea of freedom, which is why one can act, or take oneself to act, only under this idea. It constitutes, as it were, the form of the thought of oneself as a rational agent.

—Henry A. Allison, "We Can Act Only under the Idea of Freedom"

The account I have sketched of the art of self-making shows it to include an unsettling amount of unconscious or subliminal manipulation along with the exercise of "pure reason." Doesn't this process itself undermine the concept of a responsible self? This question has been explored at length by Alfred Mele in *Autonomous Agents* (1995). He argues that beyond mere self-control there is *autonomy*, which he contrasts with *heteronomy*, in which a self-controlled agent is nevertheless also under the (partial) control of others. He proposes a Default Responsibility Principle: If no one *else* is responsible for your being in state A, you are. This nicely cuts off the infinite regress feared by Kane; it permits us to pass the buck to brainwashers (if such there be in your past) but not to "society" in general or to the agentless environment. Only if foresighted, purposeful agents have been manipulating you for their own ends are you absolved from personal responsibility for the actions undertaken by your body; those are not your deeds but your brainwashers' deeds in such a case. Fair enough, but educators certainly design their interactions with us in order to further their own ends, in particular the end of turning us into reliable moral agents. How are we to distinguish between good education, dubious propaganda, and bad brainwashing? When are you benefiting from a little help from your friends, and when are you being taken for a ride?

Mele's term for brainwashing is "value engineering," and he speaks disparagingly of engineering that "bypasses" people's capacities for control over their mental lives (Mele 1995, p. 166–67). As we have seen in earlier chapters, self-control of our mental lives is limited and problematic in any case, so it is no surprise that we will have a problem distinguishing engineering that bypasses our capacities from engineering that exploits them in tolerable or desirable ways. To dramatize the difference between autonomy and heteronomy, Mele devises some thought experiments about minimally different agents, Ann and Beth. Suppose, to begin with, that Ann is genuinely autonomous—whatever that might involve. Lucky Ann. Then further suppose that Beth is *just* like Ann, her psychological identical twin, you might say, but has been somehow brainwashed without her knowledge into this perhaps only apparently enviable psychological state. Beth has all the same dispositions as Ann; she is exactly as open-minded, as non-obsessive, as flexible but also as resolute as Ann, but her apparent autonomy, Mele suggests, is bogus. She is like a perfect counterfeit dollar, readily exchangeable for a Coke and some change, but nevertheless importantly, morally, inauthentic.

Thought experiments that stipulate such extreme—and extremely unrealistic—conditions are notoriously likely to beguile the philosopher's imagination, and it's important to turn all the knobs, to vary all the stipulations this way and that, to see what's actually pumping the intuitions. Normally, in the real world, the reason that differences in historical background matter (in this case, Ann's education versus Beth's brainwashing) is that they carry implications about differences in disposition or character that will make for differences of behavior in the future. This is just what has been disallowed in the imagined case, but can we take this stipulation at face value? Thought experiments about brainwashing are endemic in philosophers' discussions of free will, and a routine—but seldom commented on—feature of these thought experiments is the victim's stipulated obliviousness to the intervention. Let's see what happens when we turn this knob. Suppose, with Mele (1995, p. 169), that Beth is later informed of her secret history and given a chance to request that her brainwashing be undone. If she retrospectively endorses it, does this act *count?* Is she *henceforth* an autonomous agent? Your intuitions may balk at this, since her state when she "endorses" it is so much a product of her earlier brainwashing (by hypothesis). You may wish to object that she's *been designed to endorse her own design,* surely an empty gesture on her part. Not so. Consider the difference time might make. Suppose we wait a few years before we inform her of her secret history, giving her lots of experience in the rough-and-tumble world of moral decision-making. Since Beth is exactly as open-minded, as cognitively flexible, as Ann (by hypothesis), this experience will be as potent, as valuable to her as it would be to Ann, and hence it ought to be just as capable of *ground-*

ing an endorsement in her case as in Ann's. We can pursue this line of thought further by supposing we now turn the same knob on Ann's case: We (lying) tell *her* that she is the victim of brainwashing. She reflects on this datum and decides she approves of the way she is—she ought to, after all; she's *actually* autonomous (whatever that comes to). Does her act count for any more than Beth's? I can see no grounds for it. Perhaps more to the point, you may feel a slight tug in the direction of supposing that by lying to Ann we have actually made her somewhat worse off, in the autonomy department, than she otherwise would be—supposing she believes our lie, of course! Why? Because now she is deeply misinformed about her past, whether or not she ever uses this misinformation in her decision-making. (And it is easy enough to imagine that this misinformation might hugely flavor all her subsequent thinking on moral topics.)

But recall that Beth was also drastically misinformed before we told her of her brainwashing. Wasn't she? Mele doesn't go into this, but presumably Beth's brainwashing was concealed from her; presumably part of her psychological similarity with Ann before her secret is revealed to her is an astonishingly rich set of false pseudo-memories of a fine, autonomy-guaranteeing moral education that never happened. How else could the stipulation that she is Ann's psychological twin otherwise be maintained? *what defines brainwashing?*

Might it be simply falsehood, then, and concealment that are the defining marks of brainwashing? As long as you tell people the truth (what passes for the truth at the time you tell it) and eschew efforts to mislead them, as long as you leave them in a state from which they can make at least as good an independent assessment of their predicament as before you intervened, you are educating them, not brainwashing them. The idea that one's history might make a morally important difference without making a difference to one's future competence is not supported by Mele's thought experiments after all. His parallel to the perfect counterfeit dollar is instructive in this regard. Counterfeiting matters because of its effects on the ambient beliefs and desires of the populace about the integrity of their money, but these are general effects, not the effects of specific bills. The identification and removal of *perfect* counterfeit bills from the currency pool would be a pointless project, since the difference between a genuine dollar and a perfect counterfeit is (*ex hypothesi*) an inert historical fact. The belief that *there are* lots of perfect counterfeits among the legal tender might disturb the economy by weakening confidence in the government's control of monetary policy, but there would be no point in rounding the counterfeits up and destroying them (as opposed to rounding up and destroying a lot of dollars in circulation).

Consider Ann and Beth again. If Beth learns the truth about her brainwashing, this will no doubt send unsettling reverberations through her psyche, with who knows what effects on her moral competence. But exactly the same reverberations will be sent through Ann if she is convincingly taught the same "truth" about herself. If one is impaired, so is the other. And if Ann's autonomy depends on the truth of her own beliefs about her own past, then Beth's problem is just that she has been lied to, not that she has been put into her enviable dispositional state by "value engineering." Note, by the way, what this portends for any doctrine that sets out to defend *Stop that crow!* on the grounds that people are better off not knowing the truth: "We had to destroy human autonomy in order to save it." Not an appealing policy statement.

The genuinely autonomous agent is rational, self-controlled, and not wildly misinformed. The intuitive repugnance we feel in "morality pills" and "brainwashing," in contrast to good old-fashioned moral education, is perhaps due, then, to a dim appreciation of the utter impossibility of there being any such shortcut treatments that could actually preserve the informedness, flexibility, and open-mindedness that, in our experience, depends on a sound education. I cannot see that *knowingly* taking a pill to improve one's self-control is any more subversive to one's autonomy than knowingly fostering a modest amount of self-deception about one's powers. If you can knowingly manipulate yourself in these ways, as a consenting adult, and endorse the effects both prospectively and retrospectively, this is a fairly good test of whether you can justifiably manipulate your children in the same fashion. In Garrison Keillor's mythical town of Lake Wobegon, "all the children are above average," and this happy myth makes them better off than they otherwise would be—so long as they don't become seriously delusional about it. It is certainly an improvement on believing that the white man's ice is colder.

Another perspective on autonomy has been explored at great length by philosophers in the wake of Harry Frankfurt's influential

[handwritten margin note: higher-order desires - want to want something]

essay, "Freedom of the Will and the Concept of a Person" (1971). Frankfurt articulated the idea that a person—an adult responsible agent—differs from an animal or a young child in having a more complex psychology, in particular: higher-order desires. A person can want one thing but *want* to want something else—and act on that second-order desire. Such a capacity to reflect on, and endorse or reject, the desires one discovers in oneself is not just a symptom of maturity, Frankfurt claims; it is a criterion of personhood. This intuitively appealing idea has proven remarkably resistant to formulation in a way that avoids regress or contradiction, and one relatively recent attempt, by David Velleman, usefully highlights the role of reasoning and the requirement that we not make ourselves too small: "The agent's role, according to Frankfurt, is to reflect on the motives competing for governance of his behavior, and to determine the outcome of the competition, by taking sides with some of his motives rather than others" (Velleman 1992, p. 476). How can a person *take sides* with, or against, some of his or her own motives?

Consider the difference between two Roman Catholic monks: One is ardent, endorsing his vow of celibacy, and triumphing in the strength of his will over his genetic makeup; the other is equally celibate, but views his Catholicism as an addiction. He considers himself brainwashed, a victim of alien memes, but he just can't convince himself to take the leap and abandon the principles he was taught. Certainly there are real people that fall into these two categories, on many fronts, but what does the difference primarily consist in? Both monks are strongly *motivated* by the tenets of Roman Catholicism, but one wholeheartedly identifies with his religion and the other doesn't. Identification cannot be a matter of a pearly Cartesian ego or immaterial soul accepting some memes and rejecting others; the entity that does the endorsement has to be itself some kind of complex meme–brain structure. But how can we identify some such structure as an *agent within,* capable of "taking sides," without lapsing back into Cartesian mysteries about an independent *res cogitans* that plays the role of Boss, or at least traffic cop and judge, in the swirling competition within the brain? Velleman gives us an example reminiscent of some of Daniel Wegner's experiments, in which a submerged, only partly or even entirely unconscious conspiracy of motives, reasons, recognitions, and the like shapes the action:

Suppose that I have a long-anticipated meeting with an old friend for the purpose of resolving some minor difference: but that as we talk, his offhand comments provoke me to raise my voice in progressively sharper replies, until we part in anger. Later reflection leads me to realize that accumulated grievances had crystallized in my mind, during the weeks before the meeting, into a resolution to sever our friendship over the matter at hand, and that this resolution is what gave the hurtful edge to my remarks. . . . But do I necessarily think that I made the decision or that I executed it? . . . When my desires and beliefs engendered an intention to sever the friendship, and when that intention triggered my nasty tone, they were exercising the same causal powers that they exercise in ordinary cases, and yet they were doing so without any contribution from me. (Velleman 1992, pp. 464–65)

What would be different if there *were* such a contribution? As Velleman notes, there has to be more to an agent than just a mathematical point, since

when he takes sides with some of those motives, he bolsters them with a force additional to, and hence other than, their own. . . . What mental event or state might play this role of always directing but never undergoing such scrutiny? It can only be a motive that drives practical thought itself. (pp. 476–77)

It can only be, as Kant said long ago, a respect for reason itself: "What animates practical thought is a concern for acting in accordance with reasons" (p. 478). And where does this come from? From the upbringing that engages the child in the practice of demanding and giving reasons. The role of consciousness here is precisely to move the issue into the arena of deliberation and consideration, where *over time* the reasons for and against can be considered and negotiated. But now what about those Jesuits who (are said to) say that the first seven years are enough for them to bring up a child so that he will identify with the faith? Is that indoctrination or education? I think it is a strength, not a weakness, of the position I am sketching here that it allows that both of the Catholic monks may well be right; the first may not be deluding himself in his belief that he has the necessary autonomy to endorse his decision and mean it, and the second may be right to resent his indoctrination as well, and the differences between their upbringing may be minuscule. People are amazingly complicated beings, and what works well for one may be quite harmful to another. (The same is true of Ritalin, of course; many for whom it is prescribed should definitely not be taking it.) What, then, is the important role of such a self? The self is a system that is *given* responsibility, over time, so that it can reliably be there to *take* responsibility, so that there is somebody home to answer when questions of accountability arise. Kane and the others are right to look for a place where the buck stops. They have just been looking for the wrong sort of thing.

Chapter 9   *a reasons for things*

Human culture supported the evolution of minds powerful enough to capture the reasons for things and make them our reasons. We are not perfectly rational agents, but the social arena we live in sustains processes of dynamic interaction that both require and permit the renewal and endorsement of our reasons, making us into agents that can take responsibility for our acts. Our autonomy does not depend on anything like the miraculous suspension of causation but rather on the integrity of the processes of education and mutual sharing of knowledge.

Chapter 10

The real threats to freedom are not metaphysical but political and social. As we learn more about the conditions of human decision-making, we will have to devise, and agree upon, systems of government and law that are not hostage to false myths about human nature, that are robust in the face of further scientific discovery and technological advances. Are we freer than we want to be? We now have more power than ever to create the conditions under which we and our descendants will lead our lives.

Notes on Sources and Further Reading

Don Ross has pointed out to me that Skyrms's analysis is not entirely general, but that Ken Binmore's recent (and forbiddingly mathematical) analysis in *Game Theory and the Social Contract,* Vol. 2: *Just Playing* (1998) provides an entirely general analysis.

Elbow Room (Dennett 1984) has an earlier version of my grad-ualistic bootstrapping account, in Chapter 4, "Self-made Selves." The present account supplements, and does not rescind in any way, that account.

Peter Suber's 1992 essay, "The Paradox of Liberation" (unpub-lished, but available on the Web at http://www.earlham.edu/~peters/writing/liber.htm), provided me with many insights, as well as the wonderful quotations from James Branch Cabell and Alcoholics Anonymous used as epigraphs.

See Judith Harris's *The Nurture Assumption* (1998), on the evi-dence that children are more strongly influenced by their peers than their parents, across a wide spectrum of psychological variables.

For somewhat orthogonal comments on Goffman's "presen-tation of self in everyday life," see Robert Wright's *The Moral Animal* (1994), in the chapter on deception and self-deception.

See my "Producing Future by Telling Stories" (1996C) on the role of fairy tales in building reliable agents. Victoria McGeer's work has been a main source for my comments on scaffolding. Relevant also is the huge literature on the "child's theory of mind," which is well surveyed in Astington, Harris, and Olson 1988; Baron-Cohen 1995; and Baron-Cohen, Tager-Flusberg, and Cohen 2000.

Those wanting to investigate the attractions and pitfalls of hard determinism and its kin should consult Michael Slote's "Ethics with-out Free Will" (1990); Susan Blackmore's *The Meme Machine* (1999); and Derk Pereboom's book *Living without Free Will* (2001).

For more on extreme philosophical thought experiments that demand that we take seriously such fantasies as morality pills and brain-washing-that-leaves-no-scars, see my "Cow-sharks, Magnets, and Swampman" (Dennett 1996B).

On Hume, see David Wiggins's "Natural and Artificial Virtues: A Vindication of Hume's Scheme" (1996).

Chapter 10

THE FUTURE OF
HUMAN FREEDOM

Where will it all end? There is no more potent source of anxiety about free will than the image of the physical sciences engulfing our every deed, good or bad, in the acid broth of causal explanation, nibbling away at the soul until there is nothing left to praise or blame, to honor, respect, or love. Or so it seems to many people. And so they try to erect one barrier or another, some absolutist doctrine designed to keep these corrosive ideas at bay. This is a doomed strategy, a relic from the last millennium. Thanks to our growing understanding of nature, we have learned that such bastions only postpone catastrophe, and often make it worse. If you want to live on the beach, you had better be prepared to move when the beach shifts, as beaches do, slowly but surely. Breakwaters can "save" the shoreline only by destroying some of the features that made the shoreline such a fine place to live in the first place. The wiser move is to study the situation and then agree on some guidelines about how close to the edge is too close to build a house. But times change, and policies that made sense for decades or centuries can become obsolete and need revision. It is often said that we have to work with nature, not against it, but of course this is just the rhetoric of moderation; every human artifice thwarts or redirects some trend of nature; the trick is to figure out enough about how nature's patterns are put together so that our interference in them will achieve the results we want.

Holding the Line against Creeping Exculpation

As we learn more and more about how people make up their minds, the assumptions underlying our institutions of praise and blame, pun-

fends rolling equilibrium

ishment and treatment, education and medication will have to adjust to honor the facts as we know them, for one thing is clear: Institutions and practices based on obvious falsehoods are too brittle to trust. Few people will be willing to wager their futures on a fragile myth that they themselves can see the cracks in. In fact, our attitudes on these matters have been shifting gradually over the centuries. We now uncontroversially exculpate or mitigate in many cases that our ancestors would have dealt with much more harshly. Is this progress or are we all going soft on sin? To the fearful, this revision looks like erosion, and to the hopeful it looks like growing enlightenment, but there is also a neutral perspective from which to view the process. It looks to an evolutionist like a rolling equilibrium, never quiet for long, the relatively stable outcome of a series of innovations and counter-innovations, adjustments and meta-adjustments, an arms race that generates at least one sort of progress: growing self-knowledge, growing sophistication about who we are and what we are, and what we can and cannot do. And from this self-understanding, we fashion and re-fashion our conclusions about what we ought to do.

Here is an unanswered question left over from Chapter 9: What, in fact, are the qualifications for being a genuinely culpable miscreant, and could anybody actually meet them? Nobody's perfect, and besides, a perfect *miscreant* is a concept in danger of self-contradiction, a point that has been appreciated since Socrates. Doesn't there have to be *something* amiss in anybody who sets out knowingly to do evil? How shall we draw the line between exculpatory pathology of various sorts—he didn't know, he couldn't control himself—and people who do evil "of their own free will," knowing what they are doing? If we set the threshold too high, everybody gets off the hook; if we set it too low, we end up punishing scapegoats. The various libertarian proposals aimed at this problem land wide of the target: Frankly mysterious agent causation, quantum indeterminacy in the faculty of practical reason, moral levitation performed by immaterial souls or other spectral puppeteers—at best these doctrines cajole us into diverting our attention from a difficult puzzle and fixating on a conveniently insoluble mystery. So let's return to the problem: How *do* we draw the line, and what keeps it from retreating in the face of all the pressure from science?

Imagine trying to devise an aptitude test that would measure the flexibility of mind, general knowledge, social comprehension, and impulse-control that are arguably the minimal requirements of moral agency. Such a test could operationalize the ideal implied by our tacit understanding of responsibility: Normal adults have it, and you either have it or you don't. We could design it to have a "ceiling effect": You can't get more than 100 out of 100 points, and most people get 100. (We have no legitimate interest in differences in competence above the threshold. Unimaginative Smith may not have known what he was doing quite as clearly as his accomplice, brilliant Jones, but Smith knew quite well enough to be held accountable.) The rationale for such a policy is clear and familiar, and it seems to work well in such simple applications as automobile driver's licenses. You have to be sixteen (or fifteen, or seventeen . . .) and you have to pass a test of aptitude and knowledge of the rules. Thereafter, you are given the freedom of the road and treated as equal to any other driver. Such a policy can then be adjusted as we learn more about its effects on highway safety; night-time restrictions, apprenticeship periods, exceptions for identifiable disabilities or other special circumstances can be considered in a cost-benefit trade-off between maximizing safety and maximizing freedom.

Just such a balancing process can also be discerned to be oper-ating in the debates over grounds for exculpation or mitigation of responsibility in general. As we learn more about patterns of relative disability and their effects, we discover grounds for relocating individ-uals relative to the threshold, usually but not always in the direction of exculpating some class of people heretofore seen as clearly culpable. This creates the appearance of an ever-retreating threshold, but we need to examine that appearance more dispassionately. It is quite pos-sible for us to make major revisions in our policies about whom we incarcerate and whom we treat, for instance, without any revision in our philosophical background assumptions. After all, we don't change our concepts of guilt and innocence when we discover that some indi-vidual in prison was falsely convicted. We remove that unfortunate per-son from the set of those deemed guilty, but we don't change the criterion for set membership. It is precisely because we adhere to our standard understanding of the concept of guilt that we recognize that this person is not guilty after all. Similarly, on the strength of new evi-dence a *category* of individuals could be removed from the set of those deemed responsible without any change—in particular, without any "erosion"—of our concept of moral responsibility. We would just learn that there were fewer morally responsible people in our society than we had heretofore supposed.

The anxious mantra returns: "But where will it all end?" Aren't we headed toward a 100 percent "medicalized" society in which nobody is responsible, and everybody is a victim of one unfortunate feature of their background or another (nature or nurture)? No, we are not, because there are forces—not mysterious metaphysical forces, but readily explainable social and political forces—that oppose this trend, and they are of the same sort, really, as the forces that prevent the driving age from rising to, say, thirty! People *want* to be held accountable. The benefits that accrue to one who is a citizen in good standing in a free society are so widely and deeply appreciated that there is always a potent presumption in favor of inclusion. Blame is the price we pay for credit, and we pay it gladly under most circumstances. We pay dearly, accepting punishment and public humiliation for a chance to get back in the game after we have been caught out in some transgression. And so the best strategy for holding the line against creeping exculpation is clear: Protect and enhance the value of the games one gets to play if one is a citizen in good standing. It is erosion of these benefits, not the onward march of the human and biological sciences, that would threaten the social equilibrium. (Recall the cynical slogan that accompanied the decay and ultimate collapse of the Soviet Union: They pretend to pay us and we pretend to work.)

Since there will always be strong temptations to make yourself really small, to externalize the causes of your actions and deny responsibility, the way to counteract these is to make people an offer they can't refuse: If you want to be free, you must *take* responsibility. But what about the poor slobs who just can't hold their lives together, whose ability to resist temptation is so impaired that they are well-nigh certain to live a life of transgression and punishment? Isn't this unfair to them, a coercive offer that only masquerades as a free choice? They can't really hold up their end of the bargain, and then they get punished. They make useful scapegoats, perhaps, since the example we set with them keeps vivid the anticipation of punishment that actually deters those with slightly more self-control, but isn't this obviously unjustifiable? After all, "they couldn't do otherwise." There is a sense of this well-worn phrase that belongs in this context, but it is not the sense that incompatibilists worry about, as we shall see.

take
resp. to be free

The dynamics of the process of negotiated thresholds is perhaps most visible in the extreme cases that occasionally come before the public. What should we do, for instance, about convicted pedophiles? The recidivism rate is appalling—you really can't teach these old dogs new tricks, apparently—and the harm they can do if allowed their freedom is even more appalling (Quinsey et al. 1998). There is, however, a treatment that studies have shown to be effective in endowing pedophiles with the self-control that would render them safe enough to return to society (under some further supervision): castration. A dire remedy for a dire condition. Can it be justified? Is it "cruel and unusual punishment"? It is important that many convicted pedophiles volunteer for castration, as a vastly preferable alternative to indefinite incarceration. (One hears less complaint about the cruel and unusual punishment of releasing a sex offender into a community of quite appropriately terrified and outraged citizens bent on forming vigilante groups to hound the dangerous individual out of town.) The issue is far from resolved and is complicated by many factors. Castration achieves its main effect by stopping the flow of testosterone into the body, and this can be done chemically or surgically. Chemical castration requires repeated injections and is in general reversible, but the drugs have some bad side effects; surgical castration is not readily reversible in one regard, but its main effect on behavior can be sidestepped by self-administering testosterone—if one really wants to. But why would one want to do this? (See, e.g., Prentky 1997 and Rosler and Witztum 1998.)

The symbolic effect of castration is obviously part of what makes the issue so highly charged. If the surgical removal of, say, the appendix, had as dramatic a positive effect on the self-control of those who underwent the treatment, it is hard to believe there would be as much vehemence in the opposition to this option. I know from experience that discussing this issue in this context is going to make some readers' heads swim. "He ends up advocating castration!" No, I have raised the policy as a serious alternative but expressed no opinion on its ultimate wisdom. After all, there may well be some better, and less dire, treatment just around the corner. Moreover, suppose for the sake of argument that the recidivism rate for pedophiles is 50 percent (not far off the mark), and suppose that many pedophiles voluntarily undergo castration as the price they are willing to pay for freedom.

Roughly half of those will be "unnecessary" castrations: They wouldn't
have re-offended in any case. The problem is that we can't identify
them (now) in advance. But presumably with growing knowledge this
will improve. What should we do in the meantime? There are compelling reasons for shunning castration, and compelling reasons for
advocating it. I am using castration as an example, and inviting readers
to reflect on how strong they find the urge to respond to such an
"unspeakable" proposal by turning off their minds and turning up the
volume on their "hearts." This is part of the problem. So sure are some
people that they are being invited onto a buttered slide to perdition that
they just can't let themselves think about such issues. Philosophers are
supposed to be above such pressures, dispassionate contemplators of
every conceivable option, insulated in their ivory towers, but that is a
myth. In fact, philosophers rather relish the role of early warning
scouts, heading off a dimly imagined catastrophe before it gets a chance
to come into focus.

    Castration is a useful example, since it exposes inconsistencies
in the thinking of advocates on both sides. There are those who eagerly
seek prescription drugs for themselves to help them keep to their diets
or control blood pressure that they cannot make themselves control
through proper exercise, while denying all such high-tech crutches to
boost or supplement the willpower of those with other temptations. If
it is rational, and responsible, for them to recognize their own weaknesses and take whatever steps are currently available to heighten their
own self-control, how can they disparage the same policies in others?
The new gastric bypass surgery that seems to be a major breakthrough
for some cases of chronic obesity caused by obsessive eating is a drastic measure, but the ambient opinion *today* in many quarters is that seriously overweight people who *resist* having the operation are being
irresponsible (Gawand 2001). This may well change as we learn more
about the long-term effects, both on the obsessive eaters and on the
surrounding society and its attitudes. Such attitudes play a powerful role
in setting the conditions in which free choices are made. For instance,
eating disorders such as bulimia and anorexia nervosa are much less
common among women in Muslim countries, in which the physical
attractiveness of women plays a muted role relative to Westernized
countries (Abed 1998). Even minor revisions of societal norms, as
Gibbard notes, can have a profound effect on how individuals think

about the choices they make, and this is a key feature distinguishing human choice from animal choice.

Suppose you have a big purple spot on your back. This is a biological feature, but probably not a very important psychological feature. Suppose instead that you have a big purple spot on your nose. This is a much greater misfortune, since although both discolorations may be physiologically harmless, the blotch on your nose will no doubt interfere profoundly with your self-image, because it affects how others see you and treat you, and how you react to that treatment, and how they react to those reactions, and so forth. A purple nose is a huge psychological handicap. Its being such a handicap is, however, something that is itself readily recognizable by many, which can lead to the endorsement of social policies, practices, and attitudes that tend to minimize, or at any rate channel, the effects. What starts out as a superficial biological feature of an organism is turned into a psychological feature and, in turn, becomes a political feature in the wider world. This sort of thing doesn't happen to any great extent in the animal world. Field ethologists routinely capture and tag the animals they study, to help with re-identification of individuals over time. Many thousands of birds have lived their lives with a colored band on one leg, and perhaps as many mammals have conducted their affairs with numbered metal tags quite visible on their ears, and so far as anyone can tell, these markers do not interfere seriously with their lives, neither diminishing nor enhancing their opportunities. A human being who had to appear in public with a metal tag affixed to one ear would have to make major adjustments in life hopes and plans, and thus there is a political dimension to any decision, self-imposed or otherwise, to display such a feature.

This sensitivity to social and political reverberations that distinguishes human agency from animal agency also provides the grounds for founding human responsibility on something more promising than quantum indeterminacy. The political negotiations out of which our current practices and presumptions about responsibility emerge have nothing to do with determinism or mechanism in general, but do concern the assessment of the inevitability—or evitability—of particular features of particular agents and types of agents. Can you teach these old dogs new tricks, or not? As we noted in Chapter 3, there is an unproblematic sense in which there can be growth in ability over time

↑ in ability over time, widen opps.

in a deterministic world, as well as a widening of opportunities and what is made of them by particular deterministic agents. Such increase in ability over time is utterly invisible to the mind-set that adopts the narrow vision of possibility enshrined in the definition of determinism: "There is at any instant exactly one physically possible future." According to that vision, in a deterministic world, at any time *t,* nothing *can do* anything other than the one thing it is determined at *t* to do, and in an indeterministic world, at any time *t,* a thing *can do* as many different things—at least two—as that brand of indeterminism allows for, presumably a deep and immutable fact of physics that could not be perturbed by changes in practices or knowledge or technology. The obvious fact that people today *can do* more than people used to be able to do disappears from sight if we understand possibility this way, and yet this fact is as important as it is obvious.

Indeed, failure to deal with the implications of *this* kind of "can" now confronts ethical theorists of every persuasion. One of the few uncontroversial propositions in ethics, deserving its own simple slogan, is "*ought* implies *can*"—you are only obligated to do something you are able to do. If you are frankly unable to do X, then it is not true that you ought to do X. It is sometimes supposed that right here we see the fundamental—and obvious—connection between free will and responsibility: Since we are responsible only for what is *in our power,* and since if determinism is true, we *can* do only whatever we are determined to do, it is never the case that we *ought* to do something else, nothing else ever being in our power. But at the same time, it is even more obvious that the explosive growth of *can-do* in recent human history is rendering obsolete many of our traditional moral notions about human obligation, quite independently of any considerations about determinism or indeterminism. The sense of "can" that has the moral import is not the sense of "can" (if there is one) that depends on indeterminism.

Suppose a competent but diseased adult asked you for assistance in putting his living body into cryogenic suspension of life pending some low-probability discovery of a cure for the disease somewhere down the road. Wouldn't that be assisted suicide? Today, arguably, it is; tomorrow it may be as obviously justifiable as assisting in the administration of anesthesia to somebody about to undergo potentially life-saving surgery. We never used to have to worry about the ethics of cloning, or omnipresent electronic surveillance, or mind-altering drugs used by athletes, or genetic enhancement of embryos, and we have never had to worry much about the prospect of effective prosthetic enhancements of the ability of human agents to control themselves, but as such innovations arise, we need to have in place an understanding of responsibility that is robust enough to accommodate them gracefully.

"Thanks, I Needed That!"

use ethics to fix what we mean

The key shift in perspective that will enable this is an inversion described by Stephen White in *The Unity of the Self* (1991, Chapter 8, "Moral Responsibility"). Don't try to use metaphysics to ground ethics, he argues; put it the other way around: Use ethics to fix what we should mean by our "metaphysical" criterion. First, show how there can be an internal justification for some agent acquiescing in his own punishment—saying, in effect, "Thanks, I needed that!"—and then use that understanding to anchor and support a reading of our pivotal phrase, *could have done otherwise:* "An agent could have done other than he or she did just in case the ascription of responsibility and blame to that agent for the action in question is justified" (p. 236). In other words, the fact that free will *is* worth wanting can be used to anchor our conception of free will in a way metaphysical myths fail to do. The basic argument is meant to cover all moral praise and blame, but we can simplify the reasoning if we focus on cases of punishment by authority ("the state") as a stand-in for the broader class of cases in which although no *crime* has been committed one individual blames another for a misdeed. In many cases in the broader class, there may be no anticipated punishment other than being scolded—or just resented, thought ill of. We can monitor the generality of the argument by shifting every now and then between a legal setting (the state vs. Jones) and a moral setting (a parent admonishing a child, for instance).

The ideal for an institution of punishment, White argues, would be that every punishment should be justified *in the eyes of the person punished.* This presupposes that agents eligible for punishment are intelligent, rational, knowledgeable enough to be competent judges of the purported justification of that punishment. Their (imagined) acquiescence in their own punishment serves as a reference or pivot point for setting the threshold. Those who are incompetent to make such a judgment are surely not competent to enjoy the freedoms of citizenship without supervision, so we don't blame them (not yet, if they are young children). Those who are competent enough to appreciate the justification, and accept it, are unproblematic instances of culpable miscreants—they say so themselves, and we have no plausible grounds for not taking them at their word. That leaves those who are apparently competent but who resist acquiescence. These are the problem cases, but they are squeezed from both sides: On the one hand, they presumably desire the status of competent citizen, with its many benefits, and on the other hand, they dread the punishment, which they can escape only by declaring themselves—or revealing themselves—to be too small. (If you make yourself really small, you can externalize virtually everything.) White notes, slyly, that even the rational psychopath will have an internal justification for supporting laws that punish psychopaths, since they protect him from other psychopaths and allow him the freedom to pursue his interests as best he can.

Whether or not such a ceremony of justification is actually performed, we can imagine the scenario. Suppose you are the culprit. The state says to you, in effect: "You erred. Tough luck, but for the good of the state you are hereby asked to undergo punishment." You hear the charges, the evidence, the verdict. Let's suppose that you are guilty as charged. (The checks and balances of the system will keep pressure on the state to make its cases well, and you are encouraged to exploit that presumption in your defense.) But now the question is whether you are responsible for the act committed. We *may* frame this as the question "Could you have done otherwise?" but we wouldn't then seek testimony from metaphysicians or quantum physicists. We would seek *specific* evidence of your competence, or extenuating circumstances. Consider, in particular, a defense that cites factors that were beyond your control, factors that were put in place many years before you were born, for instance. These are relevant only insofar as you could not have known about them. If you knew that the ground on which you were building the house had been contaminated by factory refuse a hundred years ago, or *if you should have known,* you cannot cite this as a factor beyond your control. But could you have known?

more knowledge: could have done otherwise

("Ought" implies "can.") As we come to have greater and greater powers for acquiring knowledge about the factors that play a causal role in our actions, we become increasingly liable for not knowing about factors both external (e.g., the contaminated soil) and internal (e.g., your well-understood obsession with making a quick buck—you should have done something about that!). A defense of "I could not have done otherwise" that would have passed muster in olden days is no longer acceptable. You are obliged by the prevailing attitudes of society to keep up with the latest know-how on all matters over which you wish to exercise some responsibility.

The state invites you to acquiesce in your punishment and, of course, you may not acquiesce, but if the state has done its job right, you ought to. That is, the state can offer you a reason that it can defend without blushing. If you don't get it, that's your problem. If there are lots of folks that don't get it, that's the state's problem; they have set the threshold too low, or in some other way done a bad job framing the laws. How do we handle the penumbra of cases in the real, non-ideal world of people who can't get it, or whose acquiescence is a result of brainwashing or coercion? The existence of a non-empty set of punished culprits who do not competently acquiesce in their own punishment is *inevitable,* but it is not *inevitably large.* In fact, the system of negotiated thresholds has the nice property of being adjustable *over time* to minimize the set of those misclassified. As we learn of miscarriages of justice, we consider them as grounds for revision of our policies, and when we learn of categories of individuals who fall below the currently defended threshold for self-control, we face a political question of the same sort as the question we face about whether to adjust the rules for driver's licenses. And if new technologies (surgery or drugs or treatments or prosthetic devices or educational systems or warning lights or . . .) can be effective in adjusting the abilities of those who fall short, we will confront the cost-benefit trade-off of whether the good effects outweigh the harm.

Can pedophiles do otherwise? Some can and some can't, and we should consider steps that might be taken to move more of the latter group into the former. Those who can do otherwise are those who, *if* they lapse, would insist on their *right* to be punished. And when they make this claim, we should not prejudge their presumption of competence to make it—although that will be an issue in the trial. But wouldn't the occurrence of a lapse, any lapse, show that after all, they could *not* have done otherwise—at least not on the specific occasion? No. That is an illicit return to the narrow notion of the term "can." We anchor the broader notion to our practices and *hold* such individuals responsible. In the relevant sense, they could have done otherwise. (Recall the more trivial version of this phenomenon from Chapter 3: the chess program that failed to castle but could have castled all the same—even though it operates in a deterministic world and hence would always fail to castle in *exactly* that circumstance.)

But, knowing there will almost certainly be some recidivists who do lapse, isn't this just too risky a policy to adopt? Perhaps it is, but this is a political question about how much risk we are prepared to live with, not a philosophical question about whether pedophiles have some sort of metaphysical free will after all, or even a scientific question about just what makes pedophiles do what they do. As we learn more and more about the conditions—neurochemical, social, genetic—that predispose for pedophilia (and the shifting limits of evitability of these conditions), we will surely shrink the uncertainty, and hence the risk, of releasing such people from confinement, but there will always be risk. The political question is about how much risk we are prepared to tolerate in order to maintain our freedom as a society.

For centuries we've lived by the rule that no one can be punished, or detained, for *being likely to commit a crime,* but for all that time we've been quite aware of the fact that this admirable principle has its risks. What do we do about the heretofore law-abiding citizen who approaches his intended victim with a dangerous weapon? Just when may we intervene? At what point does our fellow citizen forfeit his freedom from interference? Does he have the *right* to a first blow before we can take action against him? As we learn more and more about the probabilities, and the conditions that underlie them, there will be more and more pressure to adjust our admirable principle in the interests of public safety. Notice that we have a host of clever innovations in the law that already serve this purpose—they preserve the admirable principle by creating new crimes for people to commit on their way to their main crime. We make a law that prohibits people from carrying certain dangerous weapons in public, for instance, or that institutes the new crime of conspiracy to commit another crime. It is already a crime for people with certain medical conditions to conceal that fact when they apply for certain high-risk positions. We have ways of putting the burden of knowledge on individuals so that they can make decisions parallel to the dire choice of the pedophile. And—this is the important point—if we maintain the requirement that these innovations must pass the "Thanks, I needed that!" test, we can preserve our institution of responsibility; we can keep the specter of creeping exculpation at bay. Ask yourself: Suppose you *knew* (because of lots of good science) that you suffered from a condition that made you highly likely to injure people in some way unless you submitted to treatment Z, which would make such a calamity much more *evitable;* and suppose undergoing this treatment preserved your competence in (virtually) every way. Would you be willing to undergo the treatment? Would you be in favor of a law that made undergoing the treatment a condition of preserving your freedom? In other words, are you sure that under those conditions you would have a *right* to strike the first blow? You could say, at your trial, "I have a condition, Your Honor; it was outside my control! I couldn't do otherwise," but this would be disingenuous if you knew about the opportunity. What if such a treatment had to be undergone in childhood, before the age of informed consent? Are we prepared to consider the ethical wisdom of such preemptive interventions? What standard of evidence should we require before endorsing such a "public health" measure across the board? (We already have laws mandating inoculation, even though we know to a moral certainty that some children will have bad reactions to them and die or be disabled.) The more we know, the more we can do; the more we can do, the more obligations we face. We may yearn for the good old days when ignorance was a better excuse than it is today, but we cannot turn back the clock.

It is time to recall the plight of the hapless father from Chapter 1, who bears responsibility—doesn't he?—for the death of his child. Presumably everybody has a breaking point; those who happen to encounter their personal breaking point break! How can it be fair to hold them responsible and punish them, just because some *other* person wouldn't have broken if faced with exactly their predicament? Isn't he just the victim of bad luck? And isn't it just your good luck not to have succumbed to temptation or had your weaknesses exploited by some conspiracy of events? Yes, luck figures heavily in our lives, all the time, but since we know this, we take the precautions we deem appro-

[handwritten: take precaution to minimize effects of luck, take resp.]

priate to minimize the untoward effects of luck, and then take responsibility for whatever happens. We can note that if he makes himself really small, he can externalize this whole episode in his life, almost turning it into a bad dream, a thing that happened to him, not something he did. Or he can make himself large, and then face the much more demanding task of constructing a future self that has this terrible act of omission in its biography. It is up to him, but we may hope he gets a little help from his friends. This is indeed an opportunity for a Self-Forming Action of the sort Kane draws to our attention, and we human beings are the only species that is capable of making them, but there is no need for them to be undetermined.

Are We Freer Than We Want to Be?

Perhaps if we saw where seemingly ideal inquiry leads, we would change our minds about what makes inquiry ideal. In any case, if such a method works it must work slowly, with painstaking effort along many lines.

—Allan Gibbard, *Wise Choices, Apt Feelings*

Nicholas Maxwell (1984) defines freedom as "the capacity to achieve what is of value in a range of circumstances." I think this is about as good a short definition of freedom as could be. In particular, it appropriately leaves wide open the question of just what is of value. Our unique ability to reconsider our deepest convictions about what makes life worth living obliges us to take seriously the discovery that there is no palpable constraint on what we can consider. It is all up for grabs. To some people, this is a fearful prospect, opening the gates to nihilism and relativism, letting go of God's commandments and risking a plunge into anarchy. *Stop that crow!*

I think they should have more faith in their fellow human beings, and appreciate how amazingly subtle and adroit they are, how well equipped by nature and culture to formulate and participate in well-designed societal arrangements that maximize freedom for all. Far from being anarchic, such arrangements are—and must be—exquisitely tuned to strike a stable balance between shelter and elbow room. If we cannot achieve universality (*Homo sapiens'* chauvinistic word for species-wide acceptance), we may at least be able to aspire to what Allan Gibbard calls "parochiality over the widest parish" (Gibbard 1990, p. 315). But we may be able to achieve true universality. We've done it in other domains. The philosophers' problem is to negotiate the transition from "is" to "ought"—or, more precisely, to show how we might go beyond the "merely historical" fact that certain customs and policies have had, as a matter of fact, widespread societal endorsement, and get all the way to norms that command assent in all rational agents. Successful instances of this move are known. Bootstrapping has worked in the past, and it can work here as well. We don't need a skyhook.

Consider the curious problem of drawing a straight line. A *really* straight line. How do we do it? We use a straightedge, of course. And where did we get it? Over the centuries we refined our techniques for making straighter and straighter so-called straightedges, pitting them against each other in supervised trials and mutual adjustments that have kept raising the threshold of accuracy. We now have large machines that are accurate to within a millionth of an inch over their entire length, and we have no difficulty in using our current vantage point to appreciate the practically unattainable but readily conceivable norm of a *really* straight edge. We discovered that norm, the eternal Platonic Form of the Straight, if you like, through our creative activity. We also discovered arithmetic, and many other timeless and absolute systems of truth. As Gibbard says, we *may* not find a similar limiting point to our quest for a system of ethics, but there is no *a priori* reason that I can see to rule out the prospect, once we have the ideal in place of a free society in which free inquiry can take place. The normativity implicit in these human discoveries—or are they inventions?—is itself one of the fruits of the evolutionary processes, both genetic and cultural, that have designed us to be what we are, exploiting billions of serendipitous collisions and amplifying them, the "frozen accidents" of history, as Francis Crick has called them, into our current state. Our communal process of memetic engineering over thousands of years continues today, and this book is just part of that process. It has no Archimedean perch from which it can move the world, but it can contribute, perhaps, to the refinement of our understanding of ourselves and our circumstances.

The freedom of *thought* and *action* that is necessary for discovering truth is a precursor, as we have seen, to the more expansive ideal of political or civil freedom, a meme that spreads easily, apparently. It is much more infectious than fanaticism, thank goodness. The cat is out of the bag. There is no way that enforced ignorance can win in the long run. You can't readily uneducate people. As communications technology makes it harder and harder for leaders to shield their people from outside information, and as the economic realities of the twenty-first century make it clearer and clearer that education is the most important investment any parent can make in a child, the floodgates will open all over the world, with tumultuous effects. All the flotsam and jetsam of popular culture, all the trash and scum that accumulates in the corners of a free society, will inundate these relatively pristine regions along with the treasures of modern education, equal rights for women, better health care, workers' rights, democratic ideals, and openness to the cultures of others. As the experience in the former Soviet Union shows only too clearly, the worst features of capitalism and high-tech are among the most robust replicators in this population explosion of memes, and there will be plenty of grounds for xenophobia, Luddism, and the tempting "hygiene" of backward-looking fundamentalism.

As Jared Diamond shows in *Guns, Germs, and Steel* (1997), it was European germs that brought Western Hemisphere populations to the brink of extinction, since those people had had no history in which to develop tolerance for them. In the next century it will be our memes, both tonic and toxic, that will wreak havoc on the unprepared world. *Our* capacity to tolerate the toxic excesses of freedom cannot be assumed in others, or simply exported as one more commodity. The practically unlimited educability of any human being gives us hope of success, but designing and implementing the cultural inoculation necessary to fend off disaster, while respecting the rights of those in need of inoculation, will be an urgent task of great complexity, requiring not just better social science but also sensitivity, imagination, and courage. The field of public health expanded to include cultural health will be the greatest challenge of this century.[1]

Human Freedom Is Fragile

Whales roam the oceans, birds soar blithely overhead, and, according to an old joke, a 500-pound gorilla sits wherever it wants, but none

1. The preceding two paragraphs are drawn from Dennett 1999B.

of these creatures is free in the way human beings can be free. Human freedom is not an illusion; it is an objective phenomenon, distinct from all other biological conditions and found in only one species, us. The differences between autonomous human agents and the other assemblages of nature are visible not just from an anthropocentric perspective but also from the most objective standpoints (the plural is important) achievable. Human freedom is real—as real as language, music, and money—so it can be studied objectively from a no-nonsense, scientific point of view. But like language, music, money, and other products of society, its persistence is affected by what we believe about it. So it is not surprising that our attempts to study it dispassionately are distorted by anxiety that we will clumsily kill the specimen under the microscope.

Human freedom is younger than the species. Its most important features are only several thousand years old—an eyeblink in evolutionary history—but in that short time it has transformed the planet in ways that are as salient as such great biological transitions as the creation of an oxygen-rich atmosphere and the creation of multicellular life. Freedom had to evolve like every other feature of the biosphere, and it continues to evolve today. Freedom is real now, in some happy parts of the world, and those who love it love wisely, but it is far from inevitable, far from universal. If we understand better how freedom arose, we can do a better job of preserving it for the future, and protecting it from its many natural enemies.

Our brains have been designed by natural selection, and all the products of our brains have likewise been designed, on a much swifter timescale, by physical processes in which no exemption from causality can be discerned. How, then, can our inventions, our decisions, our sins and triumphs, be any different from the beautiful but amoral webs of the spiders? How can an apple pie, lovingly created as a gift of reconciliation, be any different, morally, from an apple, "cleverly" designed by evolution to attract a frugivore to the bargain of spreading its seeds in return for some fructose? If these are treated as rhetorical questions only, implying that only a miracle could distinguish our creations from the blind, purposeless creations of material mechanisms, we will continue to spiral around the traditional problems of free will and determinism, in a vortex of uncomprehending mystery. Human acts—acts of love and genius, as well as crimes and sins—are just too far away from the happenings in atoms, swerving randomly or not, for us to be able to see at a glance how to put them into a single coherent framework. Philosophers for thousands of years have tried to bridge the gap with a bold stroke or two, either putting science in its place or putting human pride in its place—or declaring (correctly, but unconvincingly) that the incompatibility is only apparent without going into the details. By trying to answer the questions, by sketching out the non-miraculous paths that can take us all the way from senseless atoms to freely chosen actions, we open up handholds for the imagination. The compatibility of free will and science (deterministic or indeterministic—it makes no difference) is not as inconceivable as it once seemed.

  The topics investigated in this book are not just academic puzzles, delightful conceptual riddles to be solved or curious phenomena not yet captured by good theories. Many people see them to be matters of life and death, and that makes them matters of life and death, since people's fears tend to amplify the purported implications of the different analyses and distort the arguments, making them into blunt instruments of propaganda for good or ill. The emotional resonance of the word "freedom," like that of the word "God," guarantees a partisan audience, eager to pounce on any false move, any threat, any concession. The effect is that tradition usually has a free ride, or close to it. Doctrines that are endorsed by tradition should be left unexamined if at all possible, people are inclined to think, as a matter of tactical wisdom, since it will only stir up a hornet's nest if we question them. And so traditional thinking lives on, largely unchallenged, and accretes a pearly coating of spurious invulnerability over the years.

  I have tried to show, with the help of many other thinkers, that we can and should replace these sacrosanct but brittle traditions with a more naturalistic foundation. It is scary letting go of such honored precepts as the imagined conflict between determinism and freedom, and the false security of a miracle-working Self or Soul to be the place where the buck stops. Philosophical analysis, by itself, is not enough to motivate such a drastic shift in our thinking, even when it is fundamentally correct, and perhaps the most radical feature of this book by a philosopher is the preeminence given in it to the work of non-philosophers. My point has been that philosophers, *as philosophers,* cannot claim to be doing their professional duty to their very own topics unless they pay careful attention to the thinking of psychologists such as Daniel Wegner and George Ainslie, economists such as Robert Frank, biologists such as Richard Dawkins, Jared Diamond, Edward O. Wilson, and David Sloan Wilson, and the others whose ideas have played prominent roles in this book. I am not the only philosopher who holds this view, of course. Such excellent philosophers as Jon Elster, Allan Gibbard, Philip Kitcher, Alexander Rosenberg, Don Ross, Brian Skyrms, Kim Sterelny, and Elliott Sober have pushed further than I have into these rich sources of philosophical ore, in the process clarifying both the science and the philosophy.

I have not just lavished attention on the ideas of non-philosophers; in the process I have ignored the ideas of more than a few highly regarded philosophers, sidestepping several vigorously debated controversies in my own discipline without so much as a mention. To the participants in those debates I owe an explanation. Where, some may well ask, are my refutations, my proofs, my philosophical arguments demonstrating the unsoundness of their carefully crafted analyses? I have provided a few: Austin's putt, Kane's faculty of practical reasoning, and Mele's autonomy, for instance, have come in for the sort of detailed attention philosophers expect. With regard to others, I have decided to put the burden of proof on them. It takes a certain amount of shared background assumptions to make a philosophical controversy, and I have convinced myself—not proved—that my informal tales and observations challenge some of their enabling assumptions, rendering their contests optional, however diverting to those embroiled in them. I could have said exactly how and why, but it would have taken a hundred pages or more of dense textual exegesis and argument, ending up with verdicts of false alarm, an anticlimax to be shunned. That is a risky decision on my part, since it is open to them to demonstrate that I have woefully underestimated the inevitability of their shared presuppositions, but it is a risk I am prepared to take.

My aim in this book has been to demonstrate that if we accept Darwin's "strange inversion of reasoning" we can build all the way up to the best and deepest human thought on questions of morality and meaning, ethics and freedom. Far from being an enemy of these traditional explorations, the evolutionary perspective is an indispensable ally. I have not sought to replace the voluminous work in ethics with some Darwinian *alternative,* but rather to place that work on the foundation it deserves: a realistic, naturalistic, potentially unified vision of our place in nature. Recognizing our uniqueness as reflective, communicating animals does not require any human "exceptionalism" that must shake a defiant fist at Darwin and shun the insights to be harvested from that beautifully articulated and empirically anchored system of thought. We can understand how our freedom is greater than that of other creatures, and see how this heightened capacity carries moral implications: *noblesse oblige*. We are in the best position to decide what to do next, because we have the broadest knowledge and hence the best perspective on the future. What that future holds in store for our planet is up to all of us, reasoning together.

Notes on Sources and Further Reading

Robert Kane's anthology, *The Oxford Handbook of Free Will* (2001), collects newly commissioned essays by the major contributors to the philosophical literature in recent years, and readers will get useful triangulation there on the topics covered in this book.

The complex issues of punishment and recidivism are well surveyed in Quinsey et al., *Violent Offenders: Appraising and Managing Risk* (1998), a statistically sophisticated broad overview of prediction and treatment, with particular attention to psychopaths. Among its most striking findings is that psychopaths who are given training in social sensitivity and interpersonal relations while they are incarcerated are *more* likely to commit violent crimes on release: "We speculate, then, that patients learned a great deal from the intensive program but that the psychopathic offenders put their new skills to quite unintended uses" (p. 89). Philosophers need to rethink the enabling assumptions—the oversimplifications—that they typically invoke when discussing psychopaths and other problematic culprits. As usual, a philosopher's bare imagination, untrammeled by the facts, is too blunt an instrument to be of much use on such a delicate and important set of problems.

Stephen White's *The Unity of the Self* (1991), especially Chapters 8 and 9, contains an acute and detailed analysis of some of the issues I paint here in broad brush strokes, and develops arguments that should satisfy the skeptics, especially on the need for, and soundness of, the inversion he proposes. In particular, I commend his analysis of the shortcomings of earlier philosophical attempts to deal with these issues.

A fascinating book on the history of the bootstrapping process that has yielded today's (well, the 1970s') standards of straightness and precision is Wayne Moore's *Foundations of Mechanical Accuracy* (1970).

Some readers of this book have felt the lack of an account of human creativity and authorship. This was the topic of my Presidential Address to the Eastern Division of the American Philosophical Association in December 2000 (Dennett 2000B).

The relationship between free will and political freedom is incisively investigated by Philip Pettit in *A Theory of Freedom: From the Psychology to the Politics of Agency* (2001), and by Robert Nozick, in the final chapter, "The Genealogy of Ethics," of his last book, *Invariances* (2001). The role of culture, especially political and economic organization, in maintaining and enhancing freedom is demonstrated in Amartya Sen's *Development as Freedom* (1999).

As I was putting the finishing touches on this book, I received in the mail a copy of Merlin Donald's new book, *A Mind So Rare: The Evolution of Human Consciousness* (2001). Donald makes it clear on page one that he conceives of it as an antidote of sorts to my books, *Consciousness Explained* (1991A) and *Darwin's Dangerous Idea* (1995). However, the last chapter of Donald's book, "The Triumph of Consciousness," could serve quite well as the last chapter of this book. How can this be? Because Donald, like many others, has hugely underestimated the bounty to be found in Darwin's "strange inversion of reasoning." He says in his Prologue: "This book proposes that the human mind is unlike any other on this planet, not because of its biology, which is not qualitatively unique, but because of its ability to generate and assimilate culture" (p. xiii). Exactly.

BIBLIOGRAPHY

Abed, Riadh, 1998, "The Sexual Competition Hypothesis for Eating Disorders," *British Journal of Medical Psychology,* 17:4, pp. 525–47.

Ainslie, George, 2001, *Breakdown of Will,* Cambridge: Cambridge University Press.

Akins, Kathleen, 2002, "A Question of Content," in *Daniel Dennett,* Andrew Brook and Don Ross, eds., Cambridge: Cambridge University Press, pp. 206–46.

Allison, Henry A., 1997, "We Can Act Only under the Idea of Freedom," *Proceedings of the American Philosophical Association,* 71:2, pp. 39–50.

Astington, Janet, P. L. Harris, and D.R.E. Olson, eds., 1988, *Developing Theories of Mind,* New York: Cambridge University Press.

Aunger, Robert, ed., 2000, *Darwinizing Culture: The Status of Memetics as a Science,* Oxford: Oxford University Press.

———, 2002, *The Electric Meme: A New Theory of How We Think and Communicate,* New York: Free Press.

Austin, John, 1961, "Ifs and Cans," in *Philosophical Papers,* J. O. Urmson and G. Warnock, eds., Oxford: Clarendon Press.

Avital, Eytan, and Eva Jablonka, 2000, *Animal Traditions: Behavioral Inheritance in Evolution,* Cambridge: Cambridge University Press.

Baker, Nicholson, 1996, *The Size of Thoughts: Essays and Other Lumber,* New York: Random House.

Baron-Cohen, Simon, 1995, *Mindblindness: An Essay on Autism and Theory of Mind,* Cambridge, MA: MIT Press.

Baron-Cohen, Simon, H. Tager-Flusberg, and D. Cohen, eds., 2000, *Understanding Other Minds: Perspectives from Developmental Cognitive Neuroscience,* Oxford: Oxford University Press.

Behe, Michael, 1996, *Darwin's Black Box: The Biochemical Challenge to Evolution,* New York: Free Press.

Berry, Michael, 1978, "Regular and Irregular Motion," copyright American Institute of Physics, available on his Web site: http://www.phy.bris.ac.uk/staff/berry-mv.html, ISSN 0094-243X/78/016/$1.50.

Bingham, Paul M., 1999, "Human Uniqueness: A General Theory," *Quarterly Review of Biology,* 74, pp. 133–69.

Binmore, K. G., 1998, *Game Theory and the Social Contract,* Vol. 2: *Just Playing,* Cambridge, MA: MIT Press.

Blackmore, Susan, 1999, *The Meme Machine,* Oxford: Oxford University Press.

Boone, James L., and Eric Alden Smith, 1998, "A Critique of Evolutionary Archaeology," *Current Anthropology,* 39, Supplement, pp. 104–51.

Boyd, R., and P. Richerson, 1992, "Punishment Allows the Evolution of Cooperation (or Anything Else) in Sizable Groups," *Ethology and Sociobiology,* 13, pp. 171–95.

Boyer, Pascal, 2001, *Religion Explained: The Evolutionary Origins of Religious Thought,* New York: Basic Books.

Burkert, Walter, 1996, *Creation of the Sacred: Tracks of Biology in Early Religions,* Cambridge, MA: Harvard University Press.

Cabell, James Branch, 1919, reprint 1929, *Beyond Life: Dizain des Démiurges,* R. M. McBride.

Calvin, William, 1989, *The Cerebral Symphony: Seashore Reflections on the Structure of Consciousness,* New York: Bantam.

Campbell, Donald, 1975, "On the Conflicts Between Biological and Social Evolution and Between Psychology and Moral Tradition," *American Psychologist,* Dec., pp. 1103–26.

Cartmill, Matt, 1993, *A View to a Death in the Morning: Hunting and Nature through History,* Cambridge, MA: Harvard University Press.

Chisholm, Roderick, 1964, reprint 1982, "Human Freedom and the Self," The Lindley Lecture, University of Kansas, reprinted in *Free Will,* Gary Watson, ed., Oxford: Oxford University Press.

Churchland, Patricia S., 1981, "On the Alleged Backwards Referral of Experiences and Its Relevance to the Mind-Body Problem," *Philosophy of Science,* 48, pp. 165–81.

Churchland, Paul, 1995, *The Engine of Reason, The Seat of the Soul.* Cambridge, MA: MIT Press.

Clark, Thomas, 1999, "Review of *The Volitional Brain,*" in Libet et al., pp. 271–85.

Cloak, F. T., 1975, "Is a Cultural Ethology Possible?" *Human Ecology,* 3, pp. 161–82.

Coleman, Mary, 2001, "Decisions in Action: Reasons, Motivation, and the Connection Between Them," Ph.D. dissertation, Philosophy Department, Harvard University.

Cronin, Helena, 1991, *The Ant and the Peacock: Altruism and Sexual Selection from Darwin to Today,* Cambridge: Cambridge University Press.

Darwin, Charles, 1859, *On the Origin of Species by Means of Natural Selection,* London: Murray (Harvard University Press facsimile edition).

Dawkins, Richard, 1976, 2nd ed. 1989, *The Selfish Gene,* Oxford: Oxford University Press.

———, 1982, *The Extended Phenotype: The Gene as the Unit of Selection,* San Francisco: Freeman.

———, 1993, "Viruses of the Mind," in *Dennett and his Critics,* Bo Dahlbom, ed., Oxford: Blackwell.

———, 1996, *Climbing Mount Improbable,* New York: Norton.

Dennett, Daniel C., 1975, "Why the Law of Effect Will Not Go Away," *Journal for the Theory of Social Behaviour,* 5, pp. 169–87.

———, 1978, *Brainstorms: Philosophical Essays on Mind and Psychology,* Montgomery, VT: Bradford Books.

———, 1984, *Elbow Room: The Varieties of Free Will Worth Wanting,* Cambridge, MA: MIT Press and Oxford University Press.

————, 1987, *The Intentional Stance,* Cambridge, MA: MIT Press.

————, 1990, "The Interpretation of Texts, People, and Other Artifacts," *Philosophy and Phenomenological Research,* 50, pp. 177–94.

————, 1991A, *Consciousness Explained,* Boston: Little, Brown.

————, 1991B, "Real Patterns," *Journal of Philosophy,* 88, pp. 27–51, reprinted in *Brainchildren.*

————, 1993, "Learning and Labeling" (commentary on "The Cognizer's Innards," by A. Clark and A. Karmiloff-Smith), *Mind and Language,* 8:4, pp. 540–47.

————, 1995, *Darwin's Dangerous Idea: Evolution and the Meanings of Life,* New York: Simon & Schuster.

————, 1996A, *Kinds of Minds: Toward an Understanding of Consciousness,* New York: Basic Books.

————, 1996B, "Cow-sharks, Magnets, and Swampman," *Mind & Language,* 11:1, pp. 76–77.

————, 1996C, "Producing Future by Telling Stories," in K. Ford and Z. Pylyshyn, eds., *The Robot's Dilemma Revisited: The Frame Problem in Artificial Intelligence,* Norwood, NJ: Ablex, pp. 1–7.

————, 1997A, "Appraising Grace: What Evolutionary Good Is God?" (review of *Creation of the Sacred: Tracks of Biology in Early Religions,* by Walter Burkert), *The Sciences,* Jan./Feb., pp. 39–44. (A longer version, entitled, "The Evolution of Religious Memes: Who—or What—Benefits?" with a reply by Walter Burkert, appears in *Method and Theory in the Study of Religion,* 10 (1998), pp. 115–28.

————, 1997B, "How to Do Other Things with Words," Royal Institute Conference on Philosophy of Language, Supplement to *Philosophy,* 42, John Preston, ed., Cambridge University Press, 1997, pp. 219–35.

————, 1997C, "The Case of the Tell-Tale Traces: A Mystery Solved; a Skyhook Grounded," http://ase.tufts.edu/cogstud/papers/behe.htm.

————, 1998A, *Brainchildren: Essays on Designing Minds,* Cambridge, MA: MIT Press.

————, 1998B, comment on Boone and Smith 1998, "A Critique of Evolutionary Archaeology," *Current Anthropology,* 39, Supplement, pp. 157–58.

————, 1999A, review of *Having Thought: Essays in the Metaphysics of Mind,* by John Haugeland, *Journal of Philosophy,* 96, pp. 430–35.

————, 1999B, "Protecting Public Health," in "Predictions: 30 Great Minds on the Future," *Times Higher Education Supplement,* March, pp. 74–75.

————, 2000A, "Making Tools for Thinking," in *Metarepresentations: A Multidisciplinary Perspective,* Dan Sperber, ed., Oxford: Oxford University Press.

————, 2000B, "In Darwin's Wake, Where Am I?" American Philosophical Association Eastern Division Presidential Address, published in *Proceedings and Addresses of the American Philosophical Association,* 75, Nov. 2001, pp. 13–30. Also available on http://ase.tufts.edu/cogstud.

————, 2001A, "Collision Detection, Muselot, and Scribble: Some Reflections on Creativity," in *Virtual Music,* David Cope, ed., Cambridge, MA: MIT Press.

————, 2001B, "The Evolution of Culture," *The Monist,* 84:3, pp. 305–24.

————, 2001C, "The Evolution of Evaluators," in *The Evolution of Economic Diversity,* Antonio Nicita and Ugo Pagano, eds. London: Routledge.

————, 2002A, "The New Replicators," in *Encyclopedia of Evolution,* M. Pagels, ed., Oxford: Oxford University Press.

————, 2002B, "The Baldwin Effect: A Crane, not a Skyhook," in *Evolution and Learning: The Baldwin Effect Reconsidered,* Bruce Weber and David Depew, eds., Cambridge, MA: MIT Press.

————, forthcomingA, "Altruists, Chumps, and Inconstant Pluralists" (commentary on Sober and Wilson 1998), *Philosophy and Phenomenological Research.*

————, forthcomingB, review of Avital and Jablonka 2000, *Journal of Evolutionary Biology.*

————, forthcomingC, "From Typo to Thinko," in *Evolution and Culture,* edited by Steven Levinson, Cambridge, MA: MIT Press.

————, forthcomingD, *The Science of Consciousness: Removing the Philosophical Obstacles,* 2001 Jean Nicod lectures, delivered in Paris in November 2001, Cambridge, MA: MIT Press.

Dennett, Daniel C., and Marcel Kinsbourne, 1991, "Time and the Observer: The Where and When of Consciousness in the Brain," *Behavioral and Brain Sciences,* 15, pp. 183–247.

Densmore, Shannon, and Daniel Dennett, 1999, "The Virtues of Virtual Machines," *Philosophy and Phenomenological Research,* 59, pp. 747–67.

De Waal, Frans B. M., 1996, *Good Natured: The Origins of Right and Wrong in Humans and Other Animals,* Cambridge, MA: Harvard University Press.

Diamond, Jared, 1997, *Guns, Germs, and Steel: The Fates of Human Societies,* New York: Norton.

Dickerson, Debra J., 2000, *An American Story,* New York: Pantheon.

Donald, Merlin, 2001, *A Mind So Rare: The Evolution of Human Consciousness,* New York: Norton.

Dooling, Richard, 1998, *Brain Storm,* New York: Random House.

Drescher, Gary, 1991, *Made-Up Minds: A Constructivist Approach to Artificial Intelligence,* Cambridge, MA: MIT Press.

Fischer, John Martin, and Mark Ravizza, 1998, *Responsibility and Control: A Theory of Moral Responsibility,* New York: Cambridge University Press.

Frank, Robert H., 1988, *Passions within Reason: The Strategic Role of the Emotions,* New York: Norton.

Frankfurt, Harry, 1969, "Alternative Possibilities and Moral Responsibility," *Journal of Philosophy,* 65, pp. 829–33.

———, 1971, "Freedom of the Will and the Concept of a Person," *Journal of Philosophy,* 68, pp. 5–20.

Frayn, Michael, 1999, *Headlong,* London: Faber and Faber.

French, Robert M., 1995, *The Subtlety of Sameness: A Theory and Computer Model of Analogy-Making,* Cambridge, MA: MIT Press.

Gallagher, Shaun, 1998, "The Neuronal Platonist," in conversation with Michael Gazzaniga, *Journal of Consciousness Studies,* 5:5–6, pp. 706–17.

Gawand, Atul, 2001, "The Man Who Couldn't Stop Eating," *The New Yorker,* July 9, 2001, pp. 66–75.

Gazzaniga, Michael, 1998, *The Mind's Past,* Berkeley: University of California Press.

Gibbard, Allan, 1990, *Wise Choices, Apt Feelings: A Theory of Normative Judgment,* Cambridge, MA: Harvard University Press.

Giorelli, Giulio, 1997, "Sì, abbiamo un anima. Ma è fatta di tanti piccoli robot" (interview with Daniel C. Dennett), *Corriere della Sera* (Milan), April 28, 1997.

Goffman, Erving, 1959, *The Presentation of Self in Everyday Life,* New York: Anchor Doubleday.

Goldschmidt, Tijs, 1996, *Darwin's Dreampond,* Cambridge, MA: MIT Press.

Gopnik, Adam, 1999, "Culture Vultures," *The New Yorker,* May 24, pp. 27–28.

Gould, Stephen Jay, 1978, *Ever Since Darwin,* New York: Norton.

Gray, Russell D., and F. M. Jordan, 2000, "Language Trees Support the Express-train Sequence of Austronesian Expansion," *Nature,* 405, pp. 1052–55.

Greenough, W. T., and F. R. Volkmar, 1972, "Rearing Complexity Affects Branching of Dendrites in the Visual Cortex of the Rat," *Science,* 176, pp. 1445–47.

Haig, David, 1992, "Genomic Imprinting and the Theory of Parent-Offspring Conflict," *Developmental Biology,* 3, pp. 153–60.

———, 2002, *Genomic Imprinting and Kinship,* New Brunswick, NJ: Rutgers University Press.

Haig, David, and A. Grafen, 1991, "Genetic Scrambling as a Defence against Meiotic Drive," *Journal of Theoretical Biology,* 153, pp. 531–58.

Hamilton, William D., 1996, *Narrow Roads of Gene Land,* Vol. 1. *Evolution of Social Behaviour,* Oxford: W. H. Freeman.

Hardin, Garrett, 1968, "The Tragedy of the Commons," *Science,* 162, pp. 1243–48.

Harris, Judith, 1998, *The Nurture Assumption: Why Children Turn Out the Way They Do,* New York: Touchstone (Simon & Schuster).

Hart, H.L.A., and A. M. Honoré, 1959, *Causation in the Law,* Oxford: Clarendon Press.

Haugeland, John, 1985, *Artificial Intelligence: The Very Idea,* Cambridge, MA: MIT Press.

———, 1999, *Having Thought: Essays in the Metaphysics of Mind,* Cambridge, MA: Harvard University Press.

Hofstadter, Douglas, 1997, *Le Ton Beau de Marot: In Praise of the Beauty of Language,* New York: Basic Books.

Holmes, Bob, 1998, "Irresistible Illusions," *New Scientist,* 159:2150, pp. 32–37.

Honderich, Ted, 1988, *A Theory of Determinism: The Mind, Neuroscience, and Life-Hopes,* Oxford: Oxford University Press.

Honoré, A. M., 1964, "Can and Can't," *Mind,* 73:292, pp. 463–79.

Hooper, Lora V., Lynn Bry, Per G. Falk, and Jeffrey I. Gordon, 1998, "Host-microbial Symbiosis in the Mammalian Intestine: Exploring an Internal Ecosystem," *BioEssays,* 20:4, pp. 336–43.

Hume, David, 1739, reprint 1964, *A Treatise of Human Nature,* L. A. Selby-Bigge, ed., Oxford: Clarendon Press.

James, William, 1897, reprint 1956, *The Will to Believe and Other Essays,* New York: Dover.

————, 1907, reprint 1975, *Pragmatism,* introduction by H. S. Thayer, Cambridge, MA: Harvard University Press.

Jensen, A. R., 1979, "g: Outmoded Theory or Unconquered Frontier?" *Creative Science and Technology,* 11, pp. 16–29.

Kane, Robert, 1996, *The Significance of Free Will,* Oxford: Oxford University Press.

————, 1999, "Responsibility, Luck, and Chance: Reflections on Free Will and Indeterminism," *Journal of Philosophy,* 96, pp. 217–40.

————, ed., 2001, *The Oxford Handbook of Free Will,* New York: Oxford University Press.

Kant, Immanuel, 1784, reprint 1970, "Idea for a Universal History with a Cosmopolitan Purpose," translated by H. B. Nisbet, in *Kant's Political Writings,* Hans Reiss, ed., Cambridge: Cambridge University Press.

Kass, Leon R., 1998, "Beyond Biology" (review of *Staying Human in the Genetic Future,* by Brian Appleyard), *New York Times Book Review,* Aug. 23, pp. 7–8.

Katz, Leonard D., 2000, "Toward Good and Evil: Evolutionary Approaches to Aspects of Human Morality," *Journal of Consciousness Studies,* 7:1–2. Also appears as *Evolutionary Origins of Morality: Cross-Disciplinary Perspectives,* Leonard D. Katz, ed., Bowling Green, OH: Imprint Academic, 2000.

Kornhuber, H. H., and L. Deecke, 1965, "Hirnpotentialänderungen bei Willkürbewegungen und passiven Bewegungen des Menschen: Bereitschaftspotential und reafferente Potentiale," *Pflügers Arch. ges. Physiol.,* 284, pp. 1–17.

Kripke, Saul, 1972, "Naming and Necessity," in *Semantics of Natural Language,* D. Davidson and G. Harman, eds., Dordrecht: Reidel.

Laplace, Pierre-Simon, 1814, reprint 1951, *A Philosophical Essay on Probabilities,* translated by F. W. Truscott and F. L. Emory, New York: Dover.

Leigh, E. G., 1971, *Adaptation and Diversity: Natural History and the Mathematics of Evolution,* San Francisco: Freeman, Cooper.

Lewis, David, 1973, *Counterfactuals,* Cambridge, MA: Harvard University Press.

———, 2000, "Causation as Influence," *Journal of Philosophy,* 97, pp. 182–97.

Lewontin, Richard, Steven Rose, and Leon Kamin, 1984, *Not in Our Genes: Biology, Ideology, and Human Nature,* New York: Pantheon.

Libet, Benjamin, 1981, "The Experimental Evidence for Subjective Referral of a Sensory Experience Backwards in Time: Reply to P. S. Churchland," *Philosophy of Science,* 48, pp. 182–97.

———, 1993, "The Neural Time Factor in Conscious and Unconscious Mental Events," *Experimental and Theoretical Studies of Consciousness,* Ciba Foundation Symposium #174, Chichester: Wiley.

———, 1996, "Neural Time Factors in Conscious and Unconscious Mental Function," in *Toward a Science of Consciousness,* S. R. Hameroff, A. Kaszniak, and A. Scott, eds., Cambridge, MA: MIT Press.

———, 1999, "Do We Have Free Will?" in Libet et al., 1999, pp. 45–55.

Libet, Benjamin, Anthony Freeman, and Keith Sutherland, 1999, *The Volitional Brain: Towards a Neuroscience of Free Will,* Thorverton, UK: Imprint Academic.

Libet, Benjamin, C. A. Gleason, E. W. Wright, and D. K. Pearl, 1983, "Time of Conscious Intention to Act in Relation to Onset of Cerebral Activities (Readiness Potential); the Unconscious Initiation of a Freely Voluntary Act," *Brain* 106, pp. 623–42.

MacKay, D. M., 1960, "On the Logical Indeterminacy of a Free Choice," *Mind,* 69, pp. 31–40.

MacKenzie, Robert Beverley, 1868, *The Darwinian Theory of the Transmutation of Species Examined* (published anonymously "By a Graduate of the University of Cambridge"), London: Nisbet & Co. Quoted in a review in *Athenaeum,* 2102, Feb. 8, 1868, p. 217.

Mameli, Matteo, 2002, "Learning, Evolution, and the Icing on the Cake" (review of Avital and Jablonka 2000), *Biology and Philosophy*, 17:1 pp. 141–53.

Marx, Karl, 1867, first English edition 1887, *Capital*, translated by Samuel Moore and Edward Aveling, Moscow: Progress Publishers.

Maxwell, Nicholas, 1984, *From Knowledge to Wisdom: A Revolution in the Aims and Methods of Science*, Oxford: Blackwell.

Maynard Smith, John, 1982, reprinted 1988, "Models of Cultural and Genetic Change," in his *Games, Sex and Evolution*, Hemel Hempstead, UK: Harvester.

Maynard Smith, John, and Eörs Szathmáry, 1995, *The Major Transitions in Evolution*, Oxford: Freeman.

Maynard Smith, John, and Eörs Szathmáry, 1999, *The Origins of Life: From the Birth of Life to the Origin of Language*, Oxford: Oxford University Press.

McDonald, John F., 1998, "Transposable Elements, Gene Silencing and Macroevolution," *Trends in Ecology and Evolution*, 13, pp. 94–95.

McFarland, David, 1989, "Goals, No-Goals, and Own Goals," in *Goals, No-Goals, and Own Goals: A Debate on Goal-directed and Intentional Behaviour*, Alan Montefiore and Denis Noble, eds., London: Unwin Hyman, pp. 39–57.

McGeer, Victoria, 2001, "Psycho-practice, Psycho-theory, and the Contrastive Case of Autism," *Journal of Consciousness Studies*, 8, pp. 109–32.

McGeer, Victoria, and Philip Pettit, 2002, "The Self-Regulating Mind," *Language and Communication*, 22:3, pp. 281–99.

McLaughlin, J. A., 1925, "Proximate Cause," *Harvard Law Review*, 39:149, p. 155.

Mele, Alfred, 1995, *Autonomous Agents: From Self-Control to Autonomy*, Oxford: Oxford University Press.

Metcalfe, J., and W. Mischel, 1999, "A Hot/Cool System Analysis of Delay of Gratification: Dynamics of Willpower," *Psychological Review*, 106, pp. 3–19.

Milton, Katherine, 1992, "Civilization and Its Discontents," *Natural History*, March, pp. 37–42.

Moore, G. E., 1912, *Ethics*, New York: H. Holt.

Moore, Wayne R., 1970, *Foundations of Mechanical Accuracy*, Bridgeport, CT: Moore Special Tool Co.

Moya, Andrés, and Enrique Font, eds., *Evolution: From Molecules to Ecosystems,* Oxford: Oxford University Press.

Nesse, Randolph, ed., 2001, *Evolution and the Capacity for Commitment,* New York: Russell Sage.

Nozick, Robert, 2001, *Invariances: The Structure of the Objective World,* Cambridge, MA: Harvard University Press.

Pearl, Judea, 2000, *Causality: Models, Reasoning, and Inference,* Cambridge: Cambridge University Press.

Penrose, Roger, 1989, *The Emperor's New Mind: Concerning Computers, Minds, and the Laws of Physics,* Oxford: Oxford University Press.

———, 1994, *Shadows of the Mind: A Search for the Missing Science of Consciousness,* New York: Oxford University Press

Pereboom, Derk, 2001, *Living without Free Will,* Cambridge: Cambridge University Press.

Pessin, Andrew, and Sanford Goldberg, eds., 1996, *The Twin Earth Chronicles,* Armonk, NY: M. E. Sharpe.

Pettit, Philip, 2001, *A Theory of Freedom: From the Psychology to the Politics of Agency,* Oxford: Oxford University Press.

Pinker, Steven, 1997, *How the Mind Works,* New York: Norton.

Popper, Karl, 1951, "Indeterminism in Quantum Physics and Classical Physics," *British Journal for the Philosophy of Science,* 1, pp. 179–88.

Poundstone, William, 1985, *The Recursive Universe: Cosmic Complexity and the Limits of Scientific Knowledge,* New York: Morrow.

Prentky, R. A., 1997, "Arousal Reduction in Sexual Offenders: A Review of Antiandrogen Interventions," *Sexual Abuse: A Journal of Research and Treatment,* 9, pp. 335–48.

Pynchon, Thomas, 1973, *Gravity's Rainbow,* New York: Viking.

Quine, W.V.O., 1969, "Propositional Objects," in his *Ontological Relativity and Other Essays,* New York: Columbia University Press, pp. 147–55.

Quinsey, Vernon L., Grant T. Harris, Marnie E. Rice, and Catherine A. Cormier, 1998, *Violent Offenders: Appraising and Managing Risk,* Washington, D.C.: American Psychological Association.

Raffman, Diana, 1996, "Vagueness and Context Relativity," *Philosophical Studies,* 81:2–3, pp. 175–92.

Raine, Adrian, et al., 1994, "Birth Complications Combined with Early Maternal Rejection at Age 1 Year Predispose to Violent Crime at Age 18 Years," *Archives of General Psychiatry,* 51, pp. 984–88.

Ramachandran, Vilayanur, 1998, quoted in *New Scientist,* Sept. 5, 1998, p. 35.

Rawls, John, 1971, *A Theory of Justice,* Cambridge, MA: Harvard University Press.

Ridley, Mark, 1995 (2nd ed.), *Animal Behaviour,* Boston: Blackwell Scientific Publications.

Ridley, Matt, 1996, *The Origins of Virtue,* New York: Viking.

————, 1999, *Genome: The Autobiography of a Species in 23 Chapters,* London: Fourth Estate.

Rosler, A., and E. Witztum, 1998, "Treatment of Men with Paraphilia with a Long-acting Analogue of Gonadotropin-releasing Hormone," *New England Journal of Medicine,* 338, pp. 416–22.

Ross, Don, and Paul Dumouchel, "Emotions as Strategic Signals," available at http://www.commerce.uct.ac.za/economics/staff/personalpages/dross/emote10.rtf.

Ryle, Gilbert, 1949, *The Concept of Mind,* London: Hutchinson.

Sanford, David, 1975, "Infinity and Vagueness," *Philosophical Review,* 84, pp. 520–35.

Sartre, Jean Paul, 1943, reprint 1966, *Being and Nothingness,* translated by Hazel Barnes, Philosophical Library, New York: Simon & Schuster.

Sellars, Wilfrid, 1963, "Empiricism and the Philosophy of Mind," in his *Science, Perception, and Reality,* London: Routledge & Kegan Paul, pp. 127–96.

Sen, Amartya, 1999, *Development as Freedom,* New York: Knopf.

Skyrms, Brian, 1994A, "Sex and Justice," *Journal of Philosophy,* 91, pp. 305–20.

————, 1994B, "Darwin Meets *The Logic of Decision:* Correlation in Evolutionary Game Theory," *Philosophy of Science,* 62 pp. 503–28.

————, 1996, *Evolution of the Social Contract,* New York: Cambridge University Press.

Slote, Michael, 1990, "Ethics without Free Will," *Social Theory and Practice,* 16, pp. 369–83.

Sober, Elliott, and David Sloan Wilson, 1998, *Unto Others: The Evolution and Psychology of Unselfish Behavior,* Cambridge, MA: Harvard University Press.

Sperber, Dan, ed., 2001, *The Epidemiology of Ideas,* special issue of *The Monist,* 84:3.

Sterelny, Kim, and Paul E. Griffiths, 1999, *Sex and Death: An Introduction to Philosophy of Biology,* Chicago: University of Chicago Press.

Stich, Stephen, 1996, *Deconstructing the Mind,* Oxford: Oxford University Press.

Suber, Peter, 1992, "The Paradox of Liberation," available at http://www.earlham.edu/~peters/writing/liber.htm.

Taylor, Christopher, and Daniel Dennett, 2001, "Who's Afraid of Determinism? Rethinking Causes and Possibilities," in *Oxford Handbook of Free Will,* Robert Kane, ed., New York: Oxford University Press.

Thompson, Adrian, P. Layzell, and R. S. Zebulum, 1999, "Explorations in Design Space: Unconventional Electronics Design through Artificial Evolution," *IEEE (Institute of Electrical and Electronics Engineers) Transactions on Evolutionary Computation,* 3, pp. 167–96.

Turing, Alan, 1936, "On Computable Numbers, with an Application to the *Entscheidungsproblem,*" *Proceedings of the London Mathematical Society,* 2:42, pp. 230–65.

Van Inwagen, Peter, 1983, *An Essay on Free Will,* Oxford: Clarendon Press.

Velleman, David, 1992, "What Happens When Someone Acts?" *Mind,* 101, pp. 461–81.

Wagensberg, Jorge, 2000, "Complexity versus Uncertainty: The Question of Staying Alive," *Biology and Philosophy,* 15, pp. 493–508.

Weber, Bruce, and David Depew, eds., 2002, *Evolution and Learning: The Baldwin Effect Reconsidered,* Cambridge, MA: MIT Press.

Wegner, Daniel, 2002, *The Illusion of Conscious Will,* Cambridge, MA: MIT Press.

White, Stephen L., 1991, *The Unity of the Self,* Cambridge, MA: MIT Press.

Whitehead, Alfred North, 1933, reprint 1967, *Adventures of Ideas,* New York: Macmillan.

Wiggins, David, 1996, "Natural and Artificial Virtues: A Vindication of Hume's Scheme," in *How Should One Live? Essays on the Virtues,* Roger Crisp, ed., Oxford: Clarendon Press, pp. 131–40.

Williams, George, 1988, "Reply to Comments on 'Huxley's Evolution and Ethics in Sociobiological Perspective,' " *Zygon,* 23:4, pp. 437–38.

Wolfe, Jeremy M., George A. Alvarez, and Todd S. Horowitz, 2000, "Attention Is Fast but Volition Is Slow," *Nature,* 406, p. 691.

Wolfe, Tom, 2000, *Hooking Up,* New York: Farrar, Straus & Giroux.

Wright, Robert, 1994, *The Moral Animal: The New Science of Evolutionary Psychology,* New York: Pantheon.

————, 2000, *Nonzero: The Logic of Human Destiny,* New York: Pantheon.

Zahavi, Amotz, 1987, "The Theory of Signal Selection and Some of Its Implications," in *International Symposium on Biological Evolution, Bari, 9–14 April 1985,* V. P. Delfino, ed., Bari, Italy: Adriatici Editrici, pp. 305–27.

INDEX

PENGUIN SCIENCE

THE JOURNEY OF MAN SPENCER WELLS

'Packed with important insights into our history ... Who needs literature when science is this much fun?' *Guardian*

Spencer Wells embarks on a unique voyage of discovery, travelling the world and deciphering the genetic codes of people from the Sahara Desert to Siberia. He reveals how our DNA enables us to work out where our ancestors lived and retraces their footsteps as they spread from Africa to the far corners of the earth.

ROBOT RODNEY A. BROOKS

'Brooks gives a wonderfully personal insight into the remarkable paradigm shift that AI is still undergoing' *Guardian*

We are constantly developing more and more sophisticated technology: the robot *Sojourner* explored the surface of Mars, and the prototype *Kismet* can learn to talk, recognize faces and socialize just like a child. But, Brooks says, these automata will not take over the world. Instead the distinction between humans and robots will eventually disappear as we merge with intelligent machinery

GENES, PEOPLES AND LANGUAGES
LUIGI LUCA CAVALLI-SFORZA

'It would be a slight exaggeration to say that Cavalli-Sforza studies everything about everybody, because actually he is "only" interested in what genes, languages, archaeology, and culture can teach us about the history and migrations of everybody for the last several hundred thousand years' Jared Diamond

SIX EASY PIECES RICHARD P. FEYNMAN

Drawn from Richard Feynman's celebrated and landmark text 'Lecture on Physics', this collection of essays reveals Feynman's distinctive style while introducing the essentials of physics to the general reader.

'A delightful volume – it serves as both a primer on physics for non-scientists and as a primer on Feynman himself' Paul Davies

---

PENGUIN SCIENCE

THE ORIGIN OF LIFE PAUL DAVIES

Paul Davies presents evidence that life began billions of years ago, arguing that it may well have started on Mars and spread to Earth in rocks blasted off the Red Planet by asteroid impacts. This solution to the riddle of life's origin has sweeping implications for the nature of the universe and our place within it, and opens the way to a radical rethinking of where we came from.
'The best science writer on either side of the Atlantic' *Washington Times*

THE BLIND WATCHMAKER RICHARD DAWKINS

Science is how we know what we are, where we are and why we are. The title of this work refers to the Rev. William Paley's 1802 work, *Natural Theology*, which argued that just as finding a watch would lead you to conclude that a watchmaker must exist, so the complexity of living organisms proves that a Creator exists. Not so, says Dawkins: 'The only watchmaker in nature is the blind forces of physics, deployed in a very special way . . . it is the blind watchmaker.'

UNWEAVING THE RAINBOW RICHARD DAWKINS

Why do poets and artists so often disparage science in their work? Why does so much scientific literature compare poorly with, say, the phone book? Richard Dawkins has taken a wide-ranging view of the subjects of meaning and beauty in this examination of science, mysticism and human nature.
'the product of a beguiling and fascinating mind and one generous enough to attempt to include all willing readers in its brilliantly informed enthusiasm' Melvyn Bragg, *Observer*

SYNC STEVEN STROGATZ

'A thrilling ride – from orbital patterns to sleep cycles, from flashing fireflies to brain waves. With its contagious enthusiasm and clarity of expression, *Sync* gives us a compelling glimpse into what makes our universe tick' Brian Green
'Inspiring . . . Offers a real sense of what it's like to be at the beginning of Something Big' *New Scientist*

PENGUIN SCIENCE

DARWIN'S DANGEROUS IDEA DANIEL C. DENNETT

In one of the finest explanations of the nature and implications of Darwinian evolution ever written, Dennett moves skilfully from a firm foundation in biology to possible applications for the theory in engineering and cultural evolution, presenting an unsurpassed analysis of the objections to evolutionary theory along the way. Extremely lucid, wonderfully written, and scientifically and philosophically impeccable.

THE ORIGINS OF VIRTUE MATT RIDLEY

'Are we driven by a profoundly selfish, determinist impulse? Or is there an escape clause that enables us to be genuinely unselfish and good? In an era in which biological science is challenging traditional ethics, he has raised the debate to a new level of seriousness and importance' *Sunday Times*
'A brilliant, lucid insight into the profound implications of modern biological thinking' Bryan Appleyard

THE PRIVATE LIFE OF THE BRAIN SUSAN GREENFIELD

Drawing from neuroscience, psychology, philosophy, and everyday life, Greenfield traces the life of our mind and reveals how our childhood experiences, intense emotions like fear, depression, and euphoria, and the drugs that induce these extreme feelings dramatically affect who we are. Captivating for novices and experts alike, this intriguing book presents an enlightening journey into the human brain for anyone who has ever pondered the mystery of who we are and how the brain works.

SCIENCE: A HISTORY JOHN GRIBBIN

Filled with pioneers, visionaries, eccentrics and madmen – from Galileo to Marie Curie to Feynman and beyond, this book is an enthralling story of the men and women who changed the way we see the world.
'Essential Reading' *Sunday Times*
'A magnificent history ... enormously entertaining' *Daily Telegraph*

PENGUIN SCIENCE

SOME TIME WITH FEYNMAN LEONARD MLODINOW

In Mlodinow's intimate portrait of Feynman we look into his unfolding thoughts on the nature of creativity, his rivalry with colleague Murray Gell-Mann, his love for the women in his life, and the cancer that would ultimately kill him. The result is a fascinating picture of an irreverent, charismatic and startlingly honest man, who believed in taking risks and breaking rules and who did research not from ambition, but for the thrill of discovery.

'An accessible picture of a brilliant man' Stephen Hawking

THE BLANK SLATE STEVEN PINKER

'The best book on human nature that I or anyone else will ever read. Truly magnificent' Matt Ridley

'A passionate defence of the enduring power of human nature ... both life-affirming and deeply satisfying' Tim Lott, *Daily Telegraph*

'Brilliant ... enjoyable, informative, clear, humane' *New Scientist*

HOW THE MIND WORKS STEVEN PINKER

'Why do memories fade? Why do we lose our tempers? Why do fools fall in love? Pinker's objective in this erudite account is to explore the nature and history of the human mind' Cheryl Younson, *Sunday Times*

'Witty popular science that you enjoy reading for the writing as well as for the science' *The New York Review of Books*

THE LANGUAGE INSTINCT STEVEN PINKER

'A marvellously readable book ... illuminates every facet of human language: its biological origin, its uniqueness to humanity, its acquisition by children, its grammatical structure, the production and perception of speech, the pathology of language disorders and its unstoppable evolution' *Nature*

'An extremely valuable book, informative and well written' Noam Chomsky